Resource Efficient
LDPC Decoders

Resource Efficient LDPC Decoders
From Algorithms to Hardware Architectures

Vikram Arkalgud Chandrasetty

Principal Engineer — ASIC Design Engineering,
Western Digital Corporation, Bengaluru, India

Syed Mahfuzul Aziz

Professor — Electrical and Electronic Engineering,
University of South Australia, Adelaide, South Australia, Australia

ACADEMIC PRESS

An imprint of Elsevier

British Library Cataloguing-in-Publication Data
A catalogue record for this book is available from the British Library

Library of Congress Cataloging-in-Publication Data
A catalog record for this book is available from the Library of Congress

ISBN: 978-0-12-811255-7

For Information on all Academic Press publications
visit our website at https://www.elsevier.com/books-and-journals

Publisher: Jonathan Simpson
Acquisition Editor: Pitts, Tim
Editorial Project Manager: Kent, Charlotte
Production Project Manager: Vijayaraj Purushothaman
Cover Designer: Pearson, Victoria

Typeset by MPS Limited, Chennai, India

Contents

For additional information on the topics covered in the book, visit the companion
site: https://www.elsevier.com/books-and-journals/book-companion/9780128112557

About the Authors

Vikram Arkalgud Chandrasetty received bachelor's degree in Electronics and Communication Engineering from Bangalore University (India), master's degree in VLSI System Design from Coventry University (UK) and PhD in Computer Systems Engineering from the University of South Australia (Australia). During his postdoctoral research fellowship at the University of Newcastle (Australia), he worked on designing spatially coupled LDPC codes and hardware implementations. He reviews articles for many journals including Elsevier and IEEE Transactions. He also has substantial experience as  a professional engineer. He has worked on ASIC/FPGA design, error correction coding, electronic design automation, cryptography, and communication systems for renowned companies including Motorola and SanDisk. He is currently working on designing memory controllers for next-generation storage products in Western Digital Corporation.

Syed Mahfuzul Aziz is a professor of Electrical and Electronic Engineering at the University of South Australia. His research interests are in the areas of digital systems, integrated circuit design, wireless sensor networks and smart energy systems. He leads research teams working in low power embedded processing architectures, reconfigurable sensing platforms, integration of novel sensors with electronics and communications. Prof Aziz has extensive experience in technology applications through collaborative projects and has led many industry funded projects. His recent industry collaborations  involve emerging IoT applications in organic waste management, water and agriculture sectors. As lead investigator, he has attracted competitive funding from Australian Research Council and Australian government agencies, and also funding from various industry sectors including defence and health. Professor Aziz is a senior member of the IEEE. He was the recipient of the Prime Minister's Award for Australian University Teacher of the year in 2009.

Vikram Arkalgud Chandrasetty, PhD

Principal Engineer – ASIC Design Engineering, Western Digital Corporation,
Bengaluru, India

Syed Mahfuzul Aziz, PhD

Professor – Electrical and Electronic Engineering, University of South Australia,
Adelaide, South Australia, Australia

List of Abbreviations

ASIC	Application Specific Integrated Circuit
AWGN	Additive White Gaussian Noise
BER	Bit Error Rate
BF	Bit Flip
BMP	Bitmap
BPSK	Binary Phase Shift Keying
CCDS	Consultative Committee for space Data Systems
CD	Compact Disk
CDI	Clocks per Decoding Iteration
CN	Check Node
CNP	Check Node Processor
DC	Decode Controller
DCT	Discrete Cosine Transform
DP	Decode Processor
DSP	Digital Signal Processor
DVB	Digital Video Broadcasting
DVD	Digital Versatile Disk
EG	Euclidian Geometry
ETSI	European Telecommunications Standard Institute
FEC	Forward Error Correction
FER	Frame Error Rate
FIFO	First In First Out
FPGA	Field Programmable Gate Array
FSM	Finite State Machine
GMR	Geo Mobile Radio
GSM	Global System for Mobile communication
HD	High Definition
HDL	Hardware Description Language
HQC	Hierarchical Quasi Cyclic
IC	Integrated Circuit
IEEE	Institute of Electrical and Electronics Engineers
IMB	Intermediate Message Block
IOT	Internet of Things
IP	Intellectual Property
IR	Infrared
ITU	International Telecommunication Union
JPEG	Joint Photographic Experts Group
JRCD	Joint Row Column Decoding
LD	Layered Decoding
LDPC	Low Density Parity Check
LLR	Log-Likelihood Ratio
LP	Layered Permutation
LTE	Long Term Evolution
LUT	Look Up Table

MLC	Multi-Level Cell
MMS	Modified Min-Sum
MPEG	Moving Picture Experts Group
MS	Min-Sum
MSE	Mean Square Error
NASA	National Aeronautics and Space Administration
NGH	Next Generation broadcasting system to Handheld
NRE	Non Recurring Engineering
OTN	Optical Transport Network
OWC	Optical Wireless Communication
PAR	Placement and Routing
PCI	Peripheral Component Interconnect
PEG	Progressive Edge Growth
PLB	Processor Local Bus
PMMB	Permuted Matrix Memory Block
PSNR	Peak Signal to Noise Ratio
QAM	Quadrature Amplitude Modulation
QC	Quasi Cyclic
QKD	Quantum Key Distribution
QPSK	Quadrature Phase Shift Keying
RAM	Random Access Memory
RS	Reed-Solomon
RTL	Register Transfer Level
RTN	Random Telegraph Noise
SD	Stochastic Decoding
SDH	Synchronous Digital Hierarchy
SMP	Simplified Message Passing
SNR	Signal to Noise Ratio
SOC	System On Chip
SONET	Synchronous Optical Network
SP	Sum Product
SR	Successive Relaxation
SSD	Solid State Device
TDMP	Turbo Decoding Message Passing
UC	Unconditional Correction
UEP	Unequal Error Protection
UFS	Universal Flash Storage
USB	Universal Serial Bus
UV	Ultraviolet
VN	Variable Node
VNP	Variable Node Processor
VLSI	Very Large Scale Integrated-circuits
WBF	Weighted Bit Flip
WiMAX	Worldwide Interoperability for Microwave Access
WLAN	Wireless Local Area Network
WRAN	Wireless Regional Area Network
WSN	Wireless Sensor Network

Preface

Digital communication has become part of most applications we use today. It could be for internet access using WLAN or satellite-based Digital Video Broadcasting or even mobile applications using LTE technology. It may be possible to achieve very high communication bandwidth for these applications. But, the problem is that the reliability of information received over the communication channel is often subjected to noise. The obvious solution to this problem is incorporating error correction techniques in the communication system to correct the errors introduced during transit. One of the best performing error correction codes are Low-Density Parity-Check (LDPC) codes discovered by Gallager in 1962, but these codes only gained popularity over the last decade or so. LDPC codes can achieve excellent bit error rate (BER) performance and are very suitable for next-generation communication systems. Studies have shown that large LDPC codes can achieve BER performance very close to the Shannon Limit. Hence, these codes are of paramount interest within the research community. However, practical implementation of high performance LDPC decoders with large code lengths is a challenge faced by designers today due to the huge complexity and hardware resources required.

This book presents various LDPC decoding algorithms and resource-efficient architectures for hardware implementation. LDPC decoders are primarily based on iterative algorithms and typically consist of a large number of computational nodes, with complex interconnections among the nodes. A fully-parallel implementation of LDPC decoders with large code lengths can provide high throughput, but is costly in terms of the hardware resources required. Even today's high-end FPGAs struggle to accommodate fully-parallel LDPC decoders and do not usually leave any room for other communication circuitry. Achieving high data rates for such large LDPC decoders using a reasonable amount of hardware, yet with acceptable error correction performance, remains a challenge. The complexity of the decoding algorithms and the huge hardware requirements inhibit the use of LDPC decoders in practical communication systems.

This book presents innovative techniques to reduce the complexity of LDPC decoding algorithms without compromising the BER performance. Two low-complexity algorithms are presented, namely, simplified message passing (SMP) and modified min-sum (MMS) algorithms. Both the algorithms provide good performance at a much-reduced hardware complexity. At a BER of 10^{-5}, the SMP algorithm with "4-bit precision for variable node operation" improves the BER performance by 2.0 dB compared to the bit-flip algorithm. The MMS algorithm uses reduced (2-bit) quantisation of extrinsic messages for decoding operations. For 4-bit intrinsic messages, it suffers a small loss of 0.2 dB at a BER of 10^{-6} compared to the original min-sum algorithm. Implementation of a fully parallel MMS decoder requires 18% less slices on a Xilinx FPGA compared to a decoder based on the original "min-sum algorithm with 3-bit quantisation." The MMS decoder suffers only a negligible (0.1 dB) loss in BER performance. To further reduce hardware resource requirement, this book presents a methodology for constructing flexible LDPC codes that are suitable for designing resource-efficient partially-parallel decoder architectures. This is then followed up with the design and implementation of a memory efficient partially-parallel decoder with scalable throughput. The decoder can save up to 42% of slices and 71% of block RAMs on a Xilinx FPGA compared to other similar decoders reported in the literature. The throughput of the decoder can also be easily scaled from 55 Mbps to over 1.2 Gbps. The presented LDPC decoders have been verified and tested on FPGA for wireless application standards such as WLAN and LTE. The performance of the decoder has also been assessed by comparing the quality of transmitted and reconstructed images over a simulated communication system. The approaches, algorithms, and matrix construction techniques presented in this book provide a framework for designing resource efficient LDPC decoders for various code rates and code lengths. These architectures are, therefore, flexible and address some of the challenges associated with the practical implementation of high performance LDPC decoders.

Acknowledgements

It has been a very rewarding experience for the authors to collaborate with Elsevier on this book. A big thank you to the entire Elsevier team for getting the book from the proposal stage all the way through to the publication stage. Special thanks to the Elsevier editors, Tim Pitts and Charlotte Kent, for their outstanding commitment all the way through. The production project manager, Vijayaraj Purushothaman, and his team did an excellent job within a short time frame, so thanks to you all. The authors are grateful to the esteemed reviewers for providing valuable feedback, which helped significantly to update the content and to enhance the quality of the book. Thanks to Reza Mirza Sadeghi, Research Assistant at the University of South Australia for his assistance with collating some of the information presented in Chapter 8. Special thanks to Rachna Nagaraj for assistance with proofreading. The authors are indebted to all their colleagues with whom they have had numerous productive and at times intriguing discussions. Last but not the least, the authors express their sincere gratitude to their families for their lifelong support, without which it would have been impossible to achieve what they have achieved.

Introduction

1.1 ERROR CORRECTION IN DIGITAL COMMUNICATION SYSTEM

In a Digital Communication System, the messages generated by the source which are generally in analog form are converted to digital format and then transmitted. At the receiver end, the received digital data is converted back to analog form, which is an approximation of the original message [1]. A simple block diagram of a digital communication system is shown in Fig. 1.1.

A digital communication system consists of six basic blocks. The functional blocks at the transmitter are responsible for processing the input message, encoding, modulating, and transmitting over the communication channel. The functional blocks at the receiver perform the reverse process to retrieve the original message [2].

The aim of a digital communication system is to transmit the message efficiently over the communication channel by incorporating various data compressions (e.g., DCT, JPEG, MPEG) [3], encoding and modulation techniques, in order to reproduce the message in the receiver with the least errors. The information input, which is generally in analog form, is digitized into a binary sequence, also known as an *information sequence*. The *source encoder* is responsible for compressing the input information sequence to represent it with less redundancy. The compressed data is passed to the *channel encoder*. The channel encoder introduces some redundancy in the binary information sequence that can be used by the channel decoder at the receiver to overcome the effects of noise and interference encountered by the signal while in transit through the communication channel [4]. Hence, the redundancy added in the information message helps in increasing the reliability of the data received and also improves the fidelity of the received signal. Thus, the channel encoder aids the receiver in decoding the desired information sequence. Some of the popular channel encoders are Low Density Parity Check (LDPC) codes, Turbo codes, Convolution codes, and Reed-Solomon codes. The channel encoded data is passed to the *channel modulator*, which serves as the interface to the communication channel. The encoded sequence is modulated using suitable digital modulation techniques, i.e., Binary Phase Shift Keying (BPSK), Quadrature Phase Shift Keying (QPSK) and transmitted over the communication channel [1].

Resource Efficient LDPC Decoders. DOI: https://doi.org/10.1016/B978-0-12-811255-7.00001-0

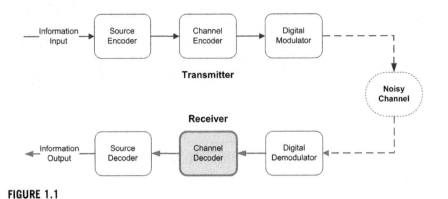

FIGURE 1.1

Block diagram of a simple Digital Communication System.

The communication channel is the physical medium used to transfer signals carrying the encoded information from the transmitter to the receiver. A range of noise and interferences can affect the information signal during transmission depending on the type of the channel medium, e.g., thermal noise, atmospheric noise, man-made noise. The communication channel can be air, wire, or optical cable [2].

At the receiver, the received modulated signal, probably incorporating some noise introduced by the channel, is demodulated by *channel demodulator* to obtain a sequence of channel encoded data in digital format. The *channel decoder* processes the received encoded sequence and decodes the message bits with the help of the redundant data inserted by the channel encoder in the transmitter. Finally, the *source decoder* reconstructs the original information message. The reconstructed information message at the receiver is probably an approximation of the original message because of errors involved in channel decoding and the distortion introduced by the source encoder and decoder [4].

1.2 FORWARD ERROR CORRECTION CODES

The importance of error correction in a digital communication system has been discussed in the previous section. All error correction codes are based on the common principle that the redundancy is added to the actual information in order to rectify any errors that could occur during transmission. In simple terms, the redundancy is appended to the information data [4]. A systematic representation of a block encoded data is shown in the Fig. 1.2. The error correction capability of the code depends on the rate of the code, which is represented by the ratio (k/n), where k is the number of information bits and n is the number of total bits including redundancy bits. The smaller the ratio, the better the error correction performance of the code.

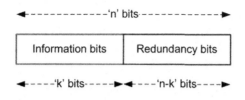

FIGURE 1.2

A systematic representation of a block encoded data.

In the channel decoding process, more errors in the *codeword* can be detected than corrected, because an error can be detected merely by the bit position. However, to correct a code, both the position and the magnitude information is required. Depending on the mechanism of adding the redundancy to the messages, error correction codes can be classified into *Convolution* codes and *Block* codes. Convolution codes' output depends on the current input as well as the previous inputs/outputs. Hence, they require memory to store additional information for encoding. In contrast, the block coding operation is memory-less, in the sense that the *codewords* are generated independently from each other. The information is processed block by block, considering each of the information bits independently. Both types of coding techniques are used in practical applications [5]. For example: Reed-Solomon block codes are widely used in CD, DVD, and computer hard drives. Hamming codes are commonly used in Random Access Memory (RAM). Turbo codes can be configured to perform as block codes and found in applications with 4G mobile telephony standard. LDPC codes are used in high speed communication standards such as DVB-S2, WLAN, and flash memory devices. Turbo and LDPC codes follow coding schemes that are decoded using iterative algorithms by exchanging soft-decision information. Consequently, these decoders can achieve bit error rate (BER) performance close to the Shannon limit [2].

LDPC codes have become very popular in recent years and have already been adopted in satellite-based Digital Video Broadcasting and Optical Communication systems [6]. With 5G expected to drive the performance of communication systems to new highs, LDPC has been adopted as an error correcting code for 5G [7]. Due to the inherent parallel nature of LDPC codes, the hardware implementation of an LDPC decoder is comparatively less complex than that of Turbo decoder with similar BER performance. Also, the LDPC decoder doesn't require designing complex inter-leaver, as the inter-leaving is distributed in the code itself. All these factors make LDPC code suitable for next generation wireless applications that require high performance encoders/decoders with low computational complexity and a low hardware resource requirement [6]. The performance and applications of LDPC codes compared to other error correction codes [8] are depicted in Fig. 1.3.

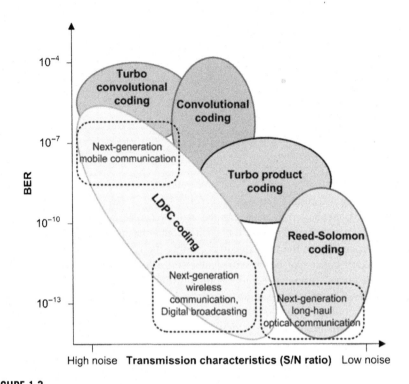

FIGURE 1.3

Performance of LDPC codes compared with other error correction codes.

REFERENCES

[1] J.G. Proakis, in: M. Salehi (Ed.), Digital Communications., fifth ed., McGraw-Hill, New York, 2008.

[2] J.C. Moreira, in: P.G. Farrell (Ed.), Essentials of Error-Control Coding, John Wiley & Sons, West Sussex, England, 2006.

[3] G.E. Blelloch, Introduction to Data Compression, Carnegie Mellon University, 2001.

[4] R.H. Morelos-Zaragoza, The Art of Error Correcting Coding, John Wiley & Sons, Chichester, 2002.

[5] J. Hagenauer, E. Offer, L. Papke, Iterative decoding of binary block and convolutional codes, IEEE Trans. Information Theory 42 (2) (1996) 429−445.

[6] G.L.L. Nicolas Fau, *LDPC (Low Density Parity Check) - A Better Coding Scheme for Wireless PHY Layers* Design and Reuse Industry Article, 2008.

[7] J. Kim, et al., Performance evaluation of reliable real-time data distribution for UAV-aided tactical networks, in: T.-h Kim, et al. (Eds.), Control and Automation, and Energy System Engineering, Springer Berlin Heidelberg, 2011, pp. 176−182.

[8] N. Tetsuo, *LDPC Adopted for Use in Comms, Broadcasting, HDDs.* Nikkei Electronics Asia, 2005.

Overview of LDPC codes

2

2.1 ORIGIN OF LDPC CODES

Low Density Parity Check (LDPC) codes were first introduced by Robert G. Gallager at MIT in his PhD thesis in 1962 [1]. The implementation complexity of LDPC codes was considered rather high. Therefore, these codes were not popular for a few decades after its invention. However, it gained popularity after the phenomenal success of Turbo Codes which were introduced in 1993 [2]. LDPC codes were formally re-introduced by MacKay and Neal in 1997 [3]. Since then, LDPC codes have become an active area of research for digital communication applications. It has been shown that LDPC codes—when optimally designed—have the capability to perform very close to the Shannon Limit [4]. The channel capacity for LDPC codes compared to other codes is shown in Fig. 2.1 [5]. The *Spectral Efficiency* curve indicates the maximum rate at which information can be transmitted over a given bandwidth. The figure presents the performance of Viterbi, LDPC, and Turbo Product Codes (TPC) at different code rates for various modulation schemes, viz. Quadrature Phase Shift Keying (QPSK), 8-Phase Shift Keying (8-PSK), and 8-Quadrature Amplitude Modulation (8-QAM). From the figure, it is clear that LDPC codes are closer to the capacity curves compared to other codes using the same modulation schemes and BER conditions, where, E_b/N_o is the ratio of energy per bit (E_b) to the noise energy (N_o). A comprehensive analysis and comparison of the code performance is available in [5].

2.2 TYPES OF LDPC CODES

Low Density Parity Check (LDPC) codes belong to a class of block codes, where the encoding is performed in blocks of data. As the name suggests, for LDPC code, the parity check matrix (H) consists of a very small number of nonzero elements. The sparseness of H determines the decoding complexity and the minimum distance of the code. Apart from the requirement that the LDPC matrix be sparse, there is no other difference between the LDPC code and any other block code. In fact, existing block codes can be used with the LDPC decoding algorithms if they can be represented by a sparse parity check matrix [6]. However,

Resource Efficient LDPC Decoders. DOI: https://doi.org/10.1016/B978-0-12-811255-7.00002-2
© 2018 Elsevier Inc. All rights reserved.

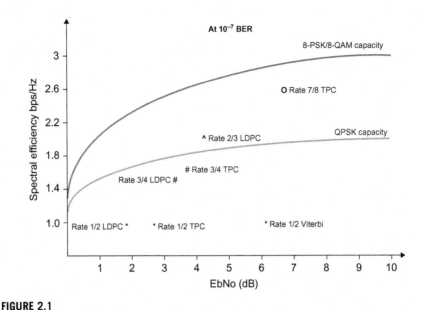

FIGURE 2.1

Comparison of channel capacity for LDPC and other error correction codes.

finding a sparse *H* matrix from existing code is difficult—or even impractical—in certain cases. Hence, LDPC codes are designed by constructing a sparse *H* matrix first and then determining the Generator matrix (G) for the code.

The LDPC matrix is described by various parameters. A *code/block length* is the sum of message bits and redundant bits. The *code rate* is defined by the ratio of the number of message bits *k* to the number of code word bits *n*. For example, one of the *code lengths* recommended in the LTE standard [7] is 576. According to this specification, a LDPC matrix (*H*) created for LTE application should have 576 columns. One of the *code rates* recommended in the LTE standard is ½. This means that half of the bits in the code word are message bits. The *H* matrix can be represented in a graph called Tanner graph. A parity check matrix with a code length of 10 bits and the corresponding Tanner graph representation of the parity-check matrix is shown in Fig. 2.2. The red squares (check nodes) represent the rows of the *H* matrix and the blue circles (variable nodes) represent the columns of the matrix. The edges connecting the nodes represent the nonzero elements in the *H* matrix.

In the *H* matrix of Figure 2.2, due to the positions of the four *non-zero elements* (*1 s*) in the first row, the first check node (c_0) has four edges connecting it to variable nodes 0, 5, 6, and 9 (v_0, v_5, v_6 and v_9). In general, a *nonzero element* in the *i*th row and *j*th column ($h_{i,j}$) represents an edge between the *i*th check node and *j*th variable node. The number of *nonzero* entries in each of the rows and

$$H = \begin{bmatrix} 1 & 0 & 0 & 0 & 0 & 1 & 1 & 0 & 0 & 1 \\ 0 & 1 & 1 & 0 & 0 & 1 & 0 & 0 & 1 & 0 \\ 1 & 0 & 0 & 0 & 1 & 0 & 0 & 0 & 1 & 1 \\ 0 & 0 & 1 & 1 & 1 & 0 & 0 & 1 & 0 & 0 \\ 0 & 1 & 0 & 1 & 0 & 0 & 1 & 1 & 0 & 0 \end{bmatrix}$$

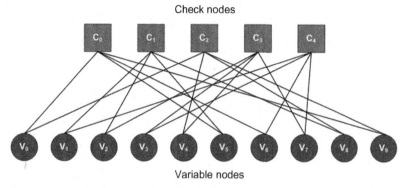

FIGURE 2.2

A parity-check matrix and its Tanner graph representation.

columns of the H matrixes collectively termed as *degree distribution*. In the figure, the check node degree is 4 and the variable node degree is 2. A cycle in the graph is a sequence of connected nodes, which start and end at the same node [6]. This cycle is known as the *Girth* of the matrix. For example, in Fig. 2.2, the nodes v_0, v_9, c_0, and c_2 form a 4-cycle girth. The *girth* in the parity check graph significantly contributes to the performance of the iterative decoding algorithms [8]. A comprehensive analysis of how the girth of a parity check matrix impacts the performance of the decoding algorithm is presented in several articles.

Since LDPC codes are decoded using iterative algorithms, construction of a sparse LDPC matrix is critically important for the overall performance of the decoder. The structure of the LDPC matrix has a direct impact on the BER performance and hardware implementation complexity. Various code construction techniques have been proposed to achieve good error correction performance and low error floor performance (An error floor is a region in the BER curve where the slope has decreased relatively to the slope at lower SNR. In an ideal decoder, the BER curve is expected to be vertical. Therefore, low error floor is expected for good performance of the LDPC decoder.) of the decoder [9]. There are different types of LDPC codes that are based on the sparseness and structure of the matrix. Some of the most popular types are briefly discussed in the following sub sections.

2.2.1 REGULAR AND IRREGULAR CODES

A parity check matrix is said to be *regular* if the degree distribution of rows and columns are uniform, otherwise the matrix is said to be *irregular*. In *regular* LDPC codes, the column and row weights are constant throughout the parity check matrix. For example, a *regular* (3, 6) parity-check matrix represents a uniform column weight of 3 and row weight of 6. A *regular* parity check matrix is preferred for hardware implementation because it leads to a constant number of edges in each of the variable and check nodes.

2.2.2 RANDOM AND PSEUDO-RANDOM CODES

A *random* sparse parity check matrix can be constructed without any constraints on the position of non-zero elements in the matrix. However, a *pseudo-random* parity-check matrix uses algorithms with certain constrains and patterns in the construction process. Both the matrixes are random in nature and will have a significant impact in hardware implementation complexity e.g., Progressive Edge Growth (PEG) is a pseudo-random algorithm for constructing LDPC codes.

2.2.3 STRUCTURED AND UNSTRUCTURED CODES

The random nature of parity check matrix construction techniques offer flexibility in design and construction process, but the lack of row and column regularity leads to complexity in hardware design. *Structured* techniques [10] can generate matrixes with well-defined pattern and degree distribution leading to relatively lower complexity in hardware implementation. However, because of such heavy constraints (degree distribution and pattern), the constructible codes are limited in terms of code parameters, e.g., Quasi-cyclic algorithms. All the parity check matrixes that do not have a well-defined pattern and structure are categorized as *unstructured*.

2.3 TERMINOLOGIES IN LDPC CODES

A variety of factors affect the design and eventually the overall performance of LDPC Codes. Hence, it is important to understand the terminologies and parameters involved in the design and development of LDPC codes.

2.3.1 LDPC CODE PARAMETERS

- *Code length (or Block length)*: The column length of the parity check matrix.
    Example: Typical code lengths for WLAN application is 648, 1296.
- *Code rate*: The ratio of actual message bits to the code length.
    Example: Typical code rates for DVB-S2 application is 1/2, 9/10.

- *Code degree/regularity*: The code is said to be *Regular* if the number of 1's in each column (d_v) and each row (d_c) are constant and is represented by (d_v, d_c). The code is *Irregular* if this condition is not true.

  Example: Typical code degree for a regular LDPC code is (3, 6) or (2, 4).
- *Girth*: The length of the shortest cycle in a given Tanner graph.

  Example: Girths can be of 4, 6, or higher.

2.3.2 SIMULATION PARAMETERS

- *Modulation*: The modulation scheme used by the transmitter, for example, BPSK, QPSK.
- *Signal to Noise Ratio (SNR)*: The ratio of energy per bit (E_b) to the noise energy (N_o). This ratio is generally represented in terms of decibels, E_b/N_o (dB).
- *Additive White Gaussian Noise (AWGN)*: commonly used to simulate background noise in a wireless communication channel.
- *Log-Likelihood Ratio (LLR)*: Logarithmic representation of bit probabilities received from the channel.
- *Quantization*: Theoretically, the LLR is a signed real number. For practical purposes, this value is quantized by the quantization function.

2.3.3 PERFORMANCE METRICS

- *Bit Error Rate (BER)*: The ratio of bits in error (B_e) to the total number of bits processed at the receiver (B_r). BER = B_e/B_r.
- *Frame Error Rate (FER)*: The ratio of frames in error (F_e) to the total number of frames processed at the receiver (F_r). FER = F_e/F_r.
- *Decoding Iterations*: The total number of iterations required for the decoder to successfully decode the given encoded data.
- *Throughput*: The total number of decoded bits a decoder is capable of generate in one second. Typically, the throughput of an LDPC decoder varies from few mega-bits-per-second to giga-bits-per-second.
- *Latency*: The delay associated in the decoder operation which is measured from the time the encoded data is fed into the system and till the time the first decoded data is available out from the system. Example: Typically measured in few milli-seconds or micro-seconds.

2.4 SUMMARY

From the above overview of LDPC codes, it can be understood that the challenges of hardware implementation of LDPC decoders are not just with the architectural design, but there is a need for low complexity decoding algorithms as well. The performance of the decoder is measured at various stages. It also depends on a number of critical factors listed in the previous sections. Achieving a balanced trade-off in terms of BER performance, logic/memory requirements and throughput, and implementation of the decoder on hardware is a challenging problem. The following chapters present some of the potential solutions to the above challenges.

REFERENCES

[1] R. Gallager, Low-density parity-check codes, IRE Trans. Information Theory 8 (1) (1962) 21−28.
[2] Berrou, C., A. Glavieux, P. Thitimajshima. Near Shannon limit error-correcting coding and decoding: Turbo-codes. In: IEEE International Conference on Communications. 1993. Geneva.
[3] D.J.C. MacKay, R.M. Neal, Near Shannon limit performance of low density parity check codes, Electronics Lett. 33 (6) (1997) 457−458.
[4] D.J.C. MacKay, Good error-correcting codes based on very sparse matrices, IEEE Trans. Information Theory 45 (2) (1999) 399−431.
[5] R. Miller, New Forward Error Correction and Modulation Technologies[cited2009 May]; Available from: <http://newerasystems.net/manuals/comtech/man-mdm-cdm600-ldpc-8qam.pdf>, 2005.
[6] S.J. Johnson, Introducing Low-Density Parity-Check Codes, University of Newcastle, Australia, 2006.
[7] T. ETSI, *136 212 LTE*. Evolved Universal Terrestrial Radio Access (EUTRA), 2016.
[8] M. Esmaeili, M. Gholami, Geometrically-structured maximum-girth LDPC block and convolutional codes, IEEE J. Select. Areas Commun. 27 (6) (2009) 831−845.
[9] Y. Lei, L. Hui, C.J.R. Shi, Code construction and FPGA implementation of a low-error-floor multi-rate low-density Parity-check code decoder, IEEE Trans. Circuits Syst. I Regular Pap. 53 (4) (2006) 892−904.
[10] G.A. Malema, Low-density parity-check codes: construction and implementation, Electrical and Electronic Engineering, The University of Adelaide, Adelaide, 2007, p. 184.

Structure and flexibility of LDPC codes

3.1 LDPC CODE CONSTRUCTION

Low density parity check (LDPC) code construction techniques are based on various algorithms. Some of the most common algorithms for generating LDPC codes are discussed in the next few sub-sections.

3.1.1 PROGRESSIVE EDGE GROWTH CODES

A parity check matrix is constructed by randomly placing *1 s* in the *all zero matrix*. However, to ensure that the LDPC matrix is regular and to obtain the largest possible *girth*, the *1 s* have to be placed in a pseudo-random fashion based on an algorithm (although the position of *1 s* appear to be randomly placed, the positions are based on an algorithm) that satisfies the above conditions. Progressive edge growth (PEG) algorithm is one such algorithm for construction of regular LDPC codes with optimum *girth* [1]. When each of the *1 s* is placed in the matrix, it is ensured that it results in the maximum desired *girth* specified to the algorithm. A Tanner graph with m check nodes and n variable nodes can be created using the H matrix. Let $S_c = \{c_0, c_1, \ldots c_{m-1}\}$ denotes the set of check nodes and $S_v = \{v_0, v_1, \ldots v_{n-1}\}$ denotes the set of variable nodes. E is the set of edges such that $E \subseteq S_c \times S_v$, with edge $(c_i, v_j) \in E$ if and only if $h_{i,j} \neq 0$, where $h_{i,j}$ denotes the entry of H at the ith row and jth column, $0 \leq i \leq m - 1$, $0 \leq j \leq n - 1$. The PEG algorithm for constructing a Tanner graph with n variable nodes and m check nodes is briefly described in Algorithm 3.1. In this algorithm, both variable nodes and check nodes are arranged according to their degrees in a nondecreasing order. So, d_{vj} is the degree of the variable node vj. N_{vj}^l and $\overline{N}_{vj}^l$ denote the set of all check nodes reached by a tree spreading from variable node vj with in the depth l, and its complement, respectively.

Resource Efficient LDPC Decoders. DOI: https://doi.org/10.1016/B978-0-12-811255-7.00003-4

ALGORITHM 3.1

1. **for** $j = 0$ to $n - 1$ **do**
2.   **for** $k = 0$ to $dvj - 1$ **do**
3.   **if** $k = 0$ **then** $E^0_{vj} \leftarrow$ edge (c_i, v_j), where E^0_{vj} is the first edge incident to v_j and c_i is a check node such that it has the lowest check-node degree under the current graph setting $E_{v0} \cup E_{v1} \cup \cdots \cup E_{vj-1}$.
4.   **else** expand a sub graph from vj up to depth l under the current graph setting such that the cardinality of N^l_{vj} stops increasing, but is less than m, or $\overline{N}^{l+1}_{vj} = \varnothing$, then $E^k_{vj} \leftarrow$ edge (c_i, v_j), where E^k_{vj} is the kth edge incident on v_j and c_i is a check node picked from $\overline{N}^l_{vj}$ having the lowest check-node degree.
5.   **end if**
6.   **end for**
7. **end for**

$$H = \begin{bmatrix} 1 & 0 & 0 & 0 & 0 & 1 & 1 & 0 & 1 & 0 \\ 0 & 1 & 1 & 0 & 0 & 0 & 0 & 0 & 1 & 1 \\ 1 & 0 & 0 & 1 & 1 & 0 & 0 & 0 & 0 & 1 \\ 0 & 0 & 1 & 0 & 1 & 0 & 1 & 1 & 0 & 0 \\ 0 & 1 & 0 & 1 & 0 & 1 & 0 & 1 & 0 & 0 \end{bmatrix}$$

FIGURE 3.1

PEG algorithm based parity check matrix structure.

It was also proved that the PEG algorithm constructs Tanner graphs having a large girth and subsequently the lower bound on the girth of the graph was proved to be [1],

$$\mathrm{gp} \geq 2 \left(\left\lfloor \frac{\log(md_c^{\max} - \frac{md_c^{\max}}{d_v^{\max}} - m + 1)}{\log[(d_v^{\max} - 1)(d_c^{\max} - 1)]} - 1 \right\rfloor + 2 \right)$$

where, $d_c^{\max}$ and $d_v^{\max}$ are the maximum degrees of the check nodes and variable nodes respectively.

A sample structure of a PEG algorithm based ½ rate (2, 4) 10-bit parity check matrix H is shown in the Fig. 3.1.

3.1.2 QUASI-CYCLIC CODES

The LDPC matrix constructed using random or pseudo-random algorithms (e.g., PEG) lack well-defined structure, which leads to complexity in hardware implementation. Therefore, structured LDPC matrices are designed to ease some of the hardware implementation complexities and other limitations imposed by unstructured matrices. Structured algorithms [2] can generate matrices with well-defined row and column regularity leading to relatively lower complexity in hardware

$$H = \begin{bmatrix} I_0 & I_1 & I_2 & . & . & I_n \\ I_1 & I_2 & . & . & . & . \\ I_2 & . & . & . & . & . \end{bmatrix} \qquad I_0 = \begin{bmatrix} 1 & 0 & 0 \\ 0 & 1 & 0 \\ 0 & 0 & 1 \end{bmatrix} \qquad I_1 = \begin{bmatrix} 0 & 1 & 0 \\ 0 & 0 & 1 \\ 1 & 0 & 0 \end{bmatrix}$$

FIGURE 3.2

Quasi-cyclic algorithm based parity check matrix structure.

implementation, however the constructible codes are limited in length, rate and girth. The quasi-cyclic algorithm is an algebraic algorithm for construction of such structured matrices.

A quasi-cyclic LDPC matrix in its elementary structure consists of a pseudo-random sparse matrix. The nonzero elements in this matrix are replaced by cyclically shifted identity matrices. The quasi-cyclic matrices are normally represented by a sparse matrix with indices that indicates the shift factor in the respective identity matrix. An example of quasi-cyclic parity check matrix H is shown in the Fig. 3.2.

3.1.3 SPATIALLY-COUPLED CODES

Spatially-coupled low density parity check (SC-LDPC) codes can be represented as a class of terminated LDPC convolutional codes. These codes have excellent sum-product decoding thresholds, especially at large code lengths [3].

An SC-LDPC code construction [4,5] starts with a protograph of a standard (d_l, d_r)-regular LDPC code, where d_l and d_r denotes the average variable node degree and check node degree. Fig. 3.3A shows a regular LDPC protograph for $d_l = 3$ and $d_r = 6$. There are $n_b = d_r/\mathbf{gcd}(d_r, d_l) = 2$ bit nodes, shown as circles and $n_c = d_l/\mathbf{gcd}(d_r, d_l) = 1$ check nodes, shown as a square, in the base protograph, where $\mathbf{gcd}$ stands for greatest common denominator. Note that when $\mathbf{gcd}(d_r, d_l) = 1$ a spatially-coupled code cannot be constructed using this method. For an SC-LDPC ensemble, a coupled chain of L of these protographs (see Fig. 3.3B) is formed by repeating the standard protograph L times and connecting it once to each of the d_l-1 protographs to its right. There are d_l-1 extra check nodes added when forming the coupled chain of protographs and this reduces the rate of the resulting spatially coupled chain when compared to the original LDPC protograph code. The rate loss diminishes as L is increased. A particular code of length N is formed from the (d_l, d_r, L) protograph chain by creating M copies of every node and edge in the coupled chain. If a particular bit node was coupled to a particular check node in the original chain, each of its copies is then connected to one of the M copies of that check node in the final code. The choice of which copy each node is connected to will depend on the construction of the code. An example of (d_l, d_r, L) SC-LDPC matrix for $M = 3$ is shown in Fig. 3.3C.

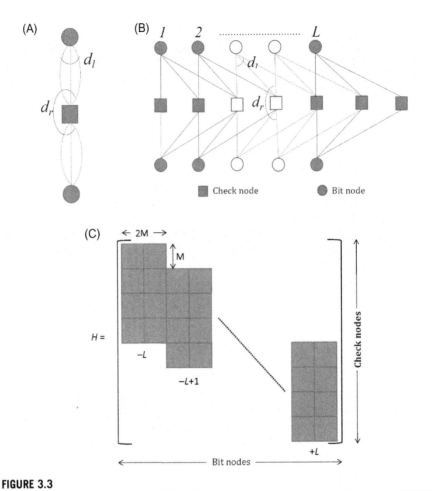

FIGURE 3.3

(A) SC-LDPC protograph, (B) SC-LDPC graph structure, and (C) SC-LDPC matrix structure.

3.1.4 REPEAT-ACCUMULATE CODES

The irregular LDPC codes are known to perform better than regular LDPC codes [6]. The repeat-accumulate (RA) codes [7] can be represented as a special subclass of irregular LDPC codes. The irregular RA codes are simple in structure with linear encoding complexity. The encoding process starts with a frame of information bits of length N. The N bits are repeated q times resulting in a frame of $q*N$ bits. A random permutation (inter-leaver) is applied to the resulting frame [8]. The permuted frame is fed to a rate-1 accumulator having a transfer function of $1/(1 + D)$. The LDPC-RA encoding process is shown in the Fig. 3.4.

$$H_2 = \begin{bmatrix} 1 & 0 & 0 & 0 & \cdots & 0 \\ 1 & 1 & 0 & 0 & \cdots & 0 \\ 0 & 1 & 1 & 0 & \cdots & 0 \\ \vdots & \vdots & \vdots & \vdots & \ddots & \vdots \\ 0 & 0 & \cdots & 1 & 1 & 0 \\ 0 & 0 & \cdots & 0 & 1 & 1 \end{bmatrix}$$

FIGURE 3.4

Repeat-accumulate parity check matrix structure.

A structured RA code can be represented as follows:

$$\mathbf{H} = [\mathbf{H_1} \ \mathbf{H_2}]$$

where,

$$\mathbf{H_1} = \text{sparse parity check matrix}$$

3.2 FLEXIBLE CODES

Quasi-cyclic (QC) based matrix construction techniques are ideal for hardware implementation because of their predictable structure and deterministic elements distribution in the matrix. However, they are less flexible for constructing matrices of variable sizes when compared to unstructured matrices. This limitation is due to the use of array of sub-matrices that are fixed in size. Construction of flexible matrices is plausible by exploiting the structured features of Hierarchical QC (HQC) methods [9,10]. It consists of permuted sub-matrices that are inserted into multiple layers of the LDPC matrix. This novel Layered Permutation (LP) technique introduces virtual randomness in the matrix similar to that of unstructured matrices, with a view to improve the decoding performance. It also uses an additional (third) level in the HQC structure to efficiently organize the structure and construct flexible matrices with variable code lengths/rates, compared to conventional HQC matrices. The following sub-sections present a detailed explanation on the construction technique and analysis of its features [11−13].

3.2.1 STRUCTURE OF THE MATRIX

In order to illustrate the matrix construction process, a ½ rate (3, 6) regular LDPC matrix is considered. A simple structure of the 3L-HQC matrix with Layered Permutation (LP) is shown in Fig. 3.5.

Level-1: The matrix has 3-Levels of hierarchy. The first level is termed as the Core matrix. This level is responsible for maintaining the rate and regularity of the LDPC matrix. For example, in the case of a ½ rate (3, 6) regular LDPC code,

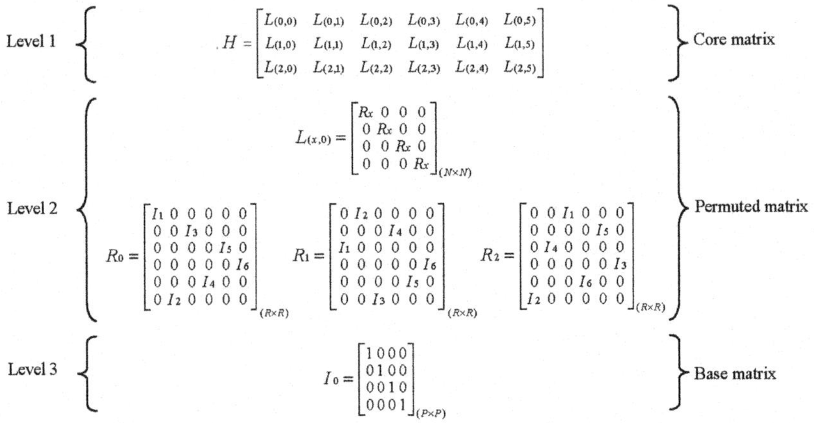

Level 1

$$H = \begin{bmatrix} L_{(0,0)} & L_{(0,1)} & L_{(0,2)} & L_{(0,3)} & L_{(0,4)} & L_{(0,5)} \\ L_{(1,0)} & L_{(1,1)} & L_{(1,2)} & L_{(1,3)} & L_{(1,4)} & L_{(1,5)} \\ L_{(2,0)} & L_{(2,1)} & L_{(2,2)} & L_{(2,3)} & L_{(2,4)} & L_{(2,5)} \end{bmatrix}$$

Core matrix

Level 2

$$L_{(x,0)} = \begin{bmatrix} R_x & 0 & 0 & 0 \\ 0 & R_x & 0 & 0 \\ 0 & 0 & R_x & 0 \\ 0 & 0 & 0 & R_x \end{bmatrix}_{(N \times N)}$$

$$R_0 = \begin{bmatrix} I_1 & 0 & 0 & 0 & 0 & 0 \\ 0 & 0 & I_3 & 0 & 0 & 0 \\ 0 & 0 & 0 & 0 & I_5 & 0 \\ 0 & 0 & 0 & 0 & 0 & I_6 \\ 0 & 0 & 0 & I_4 & 0 & 0 \\ 0 & I_2 & 0 & 0 & 0 & 0 \end{bmatrix}_{(R \times R)}$$

$$R_1 = \begin{bmatrix} 0 & I_2 & 0 & 0 & 0 & 0 \\ 0 & 0 & 0 & I_4 & 0 & 0 \\ I_1 & 0 & 0 & 0 & 0 & 0 \\ 0 & 0 & 0 & 0 & 0 & I_6 \\ 0 & 0 & 0 & 0 & I_5 & 0 \\ 0 & 0 & I_3 & 0 & 0 & 0 \end{bmatrix}_{(R \times R)}$$

$$R_2 = \begin{bmatrix} 0 & 0 & I_1 & 0 & 0 & 0 \\ 0 & 0 & 0 & 0 & I_5 & 0 \\ 0 & I_4 & 0 & 0 & 0 & 0 \\ 0 & 0 & 0 & 0 & 0 & I_3 \\ 0 & 0 & 0 & I_6 & 0 & 0 \\ I_2 & 0 & 0 & 0 & 0 & 0 \end{bmatrix}_{(R \times R)}$$

Permuted matrix

Level 3

$$I_0 = \begin{bmatrix} 1 & 0 & 0 & 0 \\ 0 & 1 & 0 & 0 \\ 0 & 0 & 1 & 0 \\ 0 & 0 & 0 & 1 \end{bmatrix}_{(P \times P)}$$

Base matrix

FIGURE 3.5

Simple structure of the 3L-HQC-LP matrix.

the Core matrix (H) consists of 3 rows and 6 columns (see Fig. 3.5). Further down the matrix construction process, each of the elements in the Core matrix that are expanded maintains a regularity of (1, 1). Hence retaining the overall regularity of (3, 6) in the complete LDPC matrix.

Level-2: The second level of the matrix is obtained by expanding each of the elements in the Core matrix with a cyclically shifted identity matrix (L) of size N. Unlike the matrix presented in [14], the matrix (L) is again expanded by placing an array of circularly shifted Permuted matrices (R_x) of size R. A Permuted matrix is constructed by placing a positive integer value randomly in the matrix R_x. Examples of integer values are shown as subscript of "I" in Fig. 3.5. This level of the matrix structure is predominantly responsible for expansion and construction of LDPC matrices with various code lengths for a particular application.

Note that different combinations of Permuted matrices are used in layers (each row of the Core matrix) of the LDPC matrix. The subscripts in each of the elements in the Core matrix (H) illustrate the layering of the Permuted matrix. For example, a subscript of (x, y) indicates that an "xth" combination of Permuted matrix is used for expansion of that particular element in the Core matrix with a cyclic shift of "y".

Level-3: In the third level, each of the nonzero elements in the Permuted matrix is expanded by a Base matrix (I). This matrix is a circularly shifted identity matrix of size P. The number of cyclic shifts in a Base matrix depends on the elements in the Permuted matrix. This is indicated by the subscript of "I" in the Permuted matrix, as shown in Fig. 3.5. The size of the Base matrix defines the number of parallel nodes (P) in the LDPC decoder. That is, the number of check nodes and variable nodes required for parallel processing.

As an example, a ½ rate (3, 6) regular LDPC code is used to demonstrate the construction of the matrix. The structure of the matrix at "Level 2" is shown in

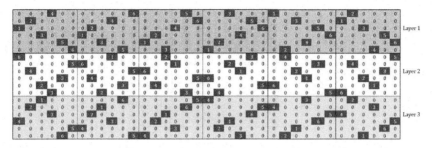

FIGURE 3.6

Intermediate structure of the 3L-HQC-LP based LDPC matrix.

Fig. 3.6. The figure clearly distinguishes the 3 layers of the permuted random matrices. Each of the permuted matrices are represented by a 6×6 square matrix. The nonzero elements in the matrix represent the magnitude of circular shift needed for an identity matrix of size $P \times P$ that is to be replaced in that position. The zero elements are also replaced by a zero matrix of same size (P). In this example, if an LDPC matrix for WLAN application is needed, then the factor P is chosen to be 18, which results in a code length of 648. Similarly, for LTE application, P is chosen to be 16 to construct a matrix for code length of 576 [15,16]. Various configurations of the matrix for different applications with multi-rate and code length is discussed in the following sections.

3.2.2 CONSTRUCTION TECHNIQUE

The 3L-HQC-LP matrix is constructed using a software model in the MATLAB environment (Available in *Appendix-1*). The technique used for construction of the matrix is illustrated in Fig 3.7. All the key parameters such as code length, code rate, and regularity of the expected LDPC matrix are provided to the software model. The model uses one of the possible combinations of matrix parameters (N, R, and P) and the permuted random matrices (R_x). The resulting LDPC matrix is verified for girth. Matrices with girth less than or equal to 4 are eliminated due to their poor decoding performance [17] and the matrix is re-constructed by varying the permuted matrix structure and circular shift parameters. When a matrix with girth greater than 4 is achieved [17], simulations are carried out to ascertain the BER, FER, and average iteration performance. Performance close/comparable to that of unstructured matrices is desired. Matrices with poor performance are rejected and the matrix re-construction procedure is repeated with different configuration of core matrix (Level 1) and permuted matrix (Level 2) until desired performance is achieved. When the desired performance is achieved, the configuration and parameters of the matrix are saved.

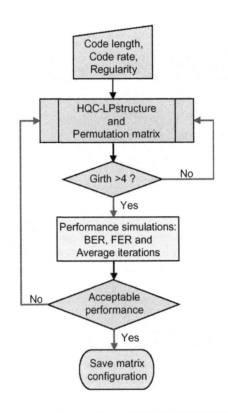

FIGURE 3.7

Illustration of the matrix construction technique.

3.2.3 STANDARD MATRIX CONFIGURATIONS

The discussed technique allows construction of LDPC matrices with different code lengths by varying N, R, and P parameters. Some of the possible configurations that are suitable for LTE [15], WLAN [18], and DVB-S2 [19] application standards are shown in Table 3.1.

In the literature, a number of decoders have been reported that use a flexible multi-rate and multi-length LDPC matrix [14,20–22]. The discussed technique also offers flexibility for constructing LDPC matrices, but without compromising the decoding performance. Table 3.1 shows examples of multirate and multilength LDPC matrices which can be constructed using the technique for multiple applications. This flexibility is achieved mainly by the layering technique and by introducing an additional level (3rd) in the hierarchy of QC matrix compared to 2-level HQC.

Table 3.1 Configurations of the Matrix for Various Applications

LTE				WLAN				DVB-S2			
P = 16				P = 18				P = 27			
CL	CR	R	N	CL	CR	R	N	CL	CR	R	N
576	1/2	1	1	648	1/2	6	1	16,200	1/3	5	20
672	1/2	7	1	1296	1/2	6	2	16,200	2/3	5	20
768	1/2	8	1	1944	1/2	6	3	64,800	1/2	8	50
864	1/2	9	1	648	2/3	6	1	64,800	1/3	8	50
960	1/2	10	1	1296	2/3	6	2	64,800	2/3	8	50
1056	1/2	11	1	1944	2/3	6	3	64,800	5/6	8	50
1152	1/2	6	2	648	5/6	6	1	–	–	–	–
1728	1/2	6	3	1296	5/6	6	2	–	–	–	–
2304	1/2	6	4	1944	5/6	6	3	–	–	–	–

Note: CL, *Code Length;* CR, *Code Rate.*

3.2.4 VISUAL ANALYSIS

The matrices constructed using the above discussed technique is verified for optimal girth. The matrices are also visually analyzed for distribution of nonzero elements. LDPC matrices constructed using PEG (unstructured random), Quasi-cyclic (QC), 3L-HQC-LP technique Spatially-coupled (SC) and Repeat-accumulate (RA) are shown in Fig. 3.8A−E.

From Fig. 3.8A it is clear that the PEG based random matrix follows a particular pattern, but the positions of the nonzero elements are not deterministic. However, the QC based matrices in Fig. 3.8B and C exhibit a particular pattern. The 3L-HQC-LP matrix has sparser distribution (Fig. 3.8B) of nonzero elements compared to 2L-HQC (Fig. 3.8B). This is due to the Layered Permutation and smaller Base matrices in the 3-level hierarchical structure. Fig. 3.8D shows the step-like structure in the matrix because of the coupling of nodes. Fig. 3.8E shows the parity check matrix along with the repeat-accumulate sequence. The visual representation of these matrices does not provide extensive details about the sparseness or layered structuring. However, this analysis is presented for additional information to provide an overview of the general structure of the matrices. The decoder performance gains achieved from the 3L-HQC-LP LDPC matrix is discussed in the next section.

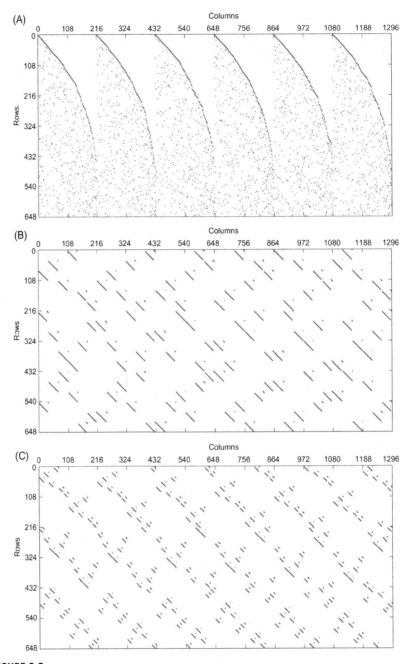

FIGURE 3.8

Visual analysis of distribution of nonzero elements in the matrix. (A) PEG based matrix
(B) Quasi-cyclic matrix (2-Level) (C) 3L-HQC with LP matrix (D) Spatially-coupled matrix
(E) Repeat-accumulate matrix.

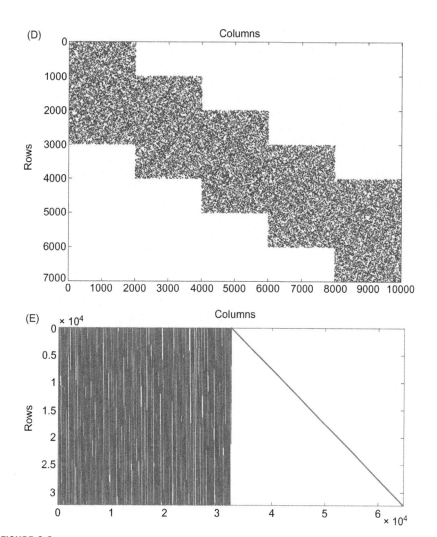

FIGURE 3.8

(Continued).

3.2.5 PERFORMANCE ANALYSIS

To analyze the decoding performance of the 3-level HQC matrices with and without layered permutation, simulations were carried out and compared against 2L-HQC and PEG based matrices. A software simulation model (Available in *Appendix-1*) has been developed using C programs and executed in the MATLAB environment [23]. A ½ rate (3, 6) regular 1296-bit LDPC decoder (complying with WLAN application standard) using Modified Min-Sum (MMS) algorithm has been used to assess the BER, FER and average iterations. For the simulations, the encoded data was assumed to have been modulated Binary Phase Shift Keying (BPSK) and passed over an Additive White Gaussian Noise (AWGN) channel. The maximum number of iterations for the algorithms was set to 10.

The BER and FER performance along with the average iterations are shown in Figs. 3.9, 3.10 and 3.11 respectively. The figures include performances of HQC matrices with and without Layered Permutation. The performance results are summarized in Table 3.2 for convenience.

Figs. 3.9−3.11 show that conventional 3L-HQC matrix has performance very close to that of 2L-HQC. The 3L-HQC offers more flexibility in generating LDPC matrices for different applications compared to 2L-HQC, and introducing the additional level (3rd level) in the matrix hierarchy has no impact on the decoding performance. The Layered Permutation technique in this chapter offers definite performance improvement. The matrix (3L-HQC-LP) outperforms both 2L-HQC and 3L-HQC by 0.3 dB at a BER of 10^{-5} and by ~ 0.35 dB at a FER of 10^{-3} (approx.). The PEG based random matrix has a marginal performance gain of less than 0.1 dB over the matrix (3L-HQC-LP) at a BER of 10^{-5} and FER of 10^{-3} (approx.). It is clear from Fig. 3.11 that, for a given E_b/N_o, the matrix with layered permutation requires fewer average iterations compared to 2L-HQC/3L-HQC.

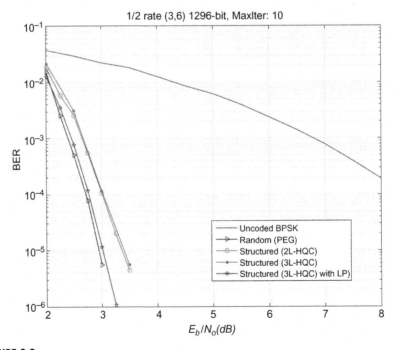

FIGURE 3.9

BER performance for various matrices.

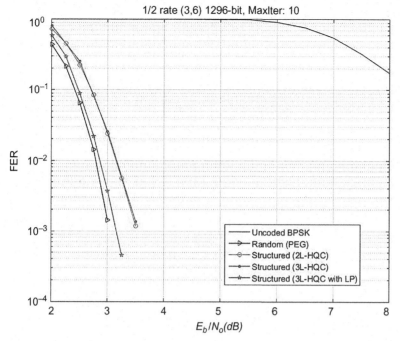

FIGURE 3.10

FER performance for various matrices.

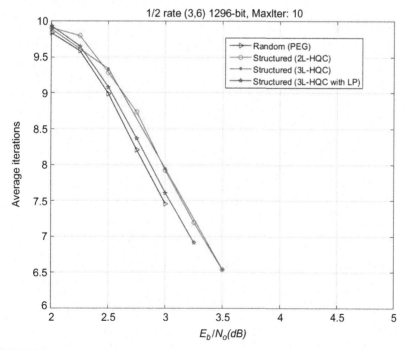

FIGURE 3.11

Average iterations for various matrices.

Table 3.2 Performance Comparison of HQC-LP Matrix with Other Matrices

Algorithms	E_b/N_o(dB) at BER of 10^{-5}	E_b/N_o(dB) at FER of 10^{-3}	Avg. Iterations at BER of 10^{-5}
PEG	2.9	3.0	7.7
2L-HQC	3.35	3.6	7.0
3L-HQC	3.4	3.6	7.0
3L-HQC with LP	3.0	3.2	7.6

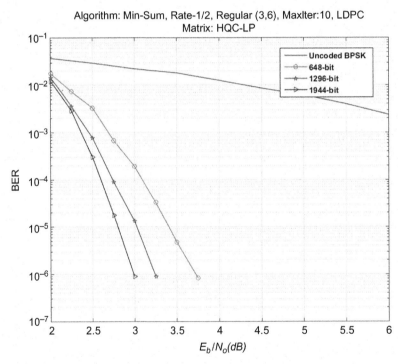

FIGURE 3.12

BER performance of 3L-HQC-LP matrix for various code lengths.

In summary, the simulation results clearly demonstrate that using Layered Permutation improves the BER performance and reduces the average iterations for decoding compared to conventional HQC based matrices.

Simulations have also been carried out to assess the decoding performance of the matrix for various code lengths complying with WLAN application. BER, FER performance, and average iterations required for various code lengths are shown in Figs. 3.12, 3.13 and 3.14 respectively. The results are also summarized

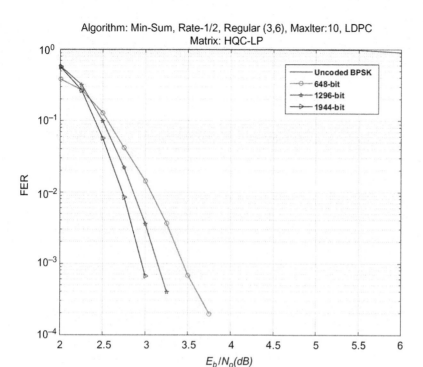

FIGURE 3.13

FER performance of 3L-HQC-LP matrices for various code lengths.

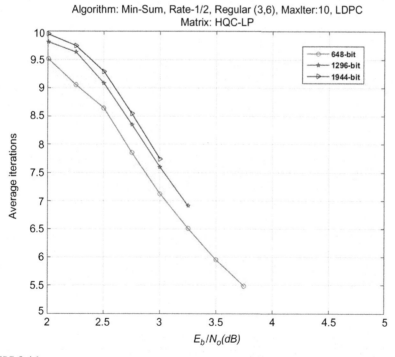

FIGURE 3.14

Average iterations of 3L-HQC-LP matrix for various code lengths.

Table 3.3 Performance Comparison of 3L-HQC-LP Matrix for Various Code Lengths

Code Lengths	E_b/N_o (dB) at BER of 10^-	E_b/N_o (dB) at FER of 10^{-3}	Avg. Iterations at BER of 10^{-6}
648	3.7	3.4	5.6
1296	3.25	3.2	6.9
1944	3.0	2.9	7.7
WLAN - IEEE 802.11 n standard			

in Table 3.3 for convenience. As expected for LDPC codes, the simulation results confirm that the BER and FER performance improve and average iterations increase for longer codes.

3.3 SUMMARY

This chapter presented a novel technique to construct LDPC matrices for different code lengths that are suitable for multiple wireless applications. The technique exploits the flexibility of matrix construction inherent in the Hierarchical Quasi-Cyclic (HQC) based approach. It is shown that the Layered Permutation (LP) technique used in the matrix helps achieving significant improvement in decoding performance compared to conventional HQC based matrices. It provides decoding performance close to Progressive Edge Growth (PEG) based matrices. The matrix offers flexibility in code construction for various application standards. These advantages of the matrix have been verified through software simulations and by comparing with other LDPC matrix structures. The flexibility of the decoder (supporting multiple code lengths, throughput and hardware requirements) is also verified by designing a prototype hardware model of partially-parallel architecture which is discussed in Chapter 6, Hardware implementation of LDPC decoders. The throughput of the decoder is easily scalable by selecting appropriate number of parallel nodes. Therefore, this technique nevertheless provides an attractive solution for implementing highly flexible LDPC decoders complying with different wireless applications.

Having constructed a custom LDPC matrix suitable for hardware implementation, the subsequent step is to design efficient partially-parallel decoder architectures. The next few chapters present implementations of resource efficient, fully-parallel, and partially-parallel decoders.

REFERENCES

[1] H. Xiao-Yu, E. Eleftheriou, D.M. Arnold. Progressive edge-growth Tanner graphs. in IEEE Global Telecommunications Conference. San Antonio, TX, 2001.

[2] G.A. Malema, Low-density parity check codes: construction and implementation, Electrical and Electronic Engineering, The University of Adelaide, Adelaide, 2007, p. 184.

[3] K. Kasai, , K. Sakaniwa. Spatially-coupled MacKay-Neal codes and Hsu-Anastasopoulos codes. in IEEE International Symposium on Information Theory Proceedings, St. Petersburg, 2011.

[4] S. Johnson, G. Lechner, Spatially coupled repeat-accumulate codes, IEEE Commun. Lett. 17 (2) (2013) 373−376.

[5] V.A. Chandrasetty, S.J. Johnson, G. Lechner, Memory-efficient quasi-cyclic spatially coupled low-density parity check and repeat-accumulate codes, IET Commun. 8 (2014) 3179−3188.

[6] X. Hua, A.H. Banihashemi, Improved progressive-edge-growth (PEG) construction of irregular LDPC codes, IEEE Commun. Lett. 8 (12) (2004) 715−717.

[7] C. Xiang, L. Huanbing, J. Hui. A class of irregular repeat accumulate code with flexible code length and without short circles. In: International Conference on Electrical and Control Engineering, Yichang, 2011.

[8] S.J. Johnson, V.A. Chandrasetty, A.M. Lance. Repeat-accumulate codes for reconciliation in continuous variable quantum key distribution. in 2016 Australian Communications Theory Workshop (AusCTW), 2016.

[9] C. Yi-Hsing, K. Mong-Kai. A High Throughput H-QC LDPC Decoder. in IEEE International Symposium on Circuits and Systems. New Orleans, LA, 2007.

[10] M. Fossorier, Quasi-cyclic low-density parity check codes from circulant permutation matrices, IEEE Trans. Information Theory 50 (8) (2004) 1788−1793.

[11] V.A. Chandrasetty, S.M. Aziz. Construction of a multi-level hierarchical quasi-cyclic matrix with layered permutation for partially-parallel LDPC decoders. in 13th International Conference on Computers and Information Technology. Dhaka, 2010.

[12] V.A. Chandrasetty, S.M. Aziz. A multi-level hierarchical quasi-cyclic matrix for implementation of flexible partially-parallel LDPC decoders. In: IEEE International Conference on Multimedia and Expo. Barcelona, Spain, 2011.

[13] V.A. Chandrasetty, S.M. Aziz, A highly flexible LDPC decoder using hierarchical quasi-cyclic matrix with layered permutation, Journal of Networks, Academy Publisher 7 (3) (2012) 441−449.

[14] N. Bonello, S. Chen, L. Hanzo. Multilevel Structured Low-Density Parity check Codes for AWGN and Rayleigh Channels. in IEEE International Conference on Communications. Beijing, 2010.

[15] ETSI, T., 136 212 LTE. Evolved Universal Terrestrial Radio Access (EUTRA), 2016.

[16] I. Bhurtah, P.C. Catherine, K.M.S. Soyjaudah. Enhancing the error-correcting performance of LDPC codes for LTE and WiFi. in International Conference on Computing, Communication & Automation, 2015.

[17] W. Xiaofu, Y. Xiaohu, Z. Chunming, A necessary and sufficient condition for determining the girth of quasi-cyclic LDPC codes, IEEE Trans. Commun. 56 (6) (2008) 854–857.

[18] 802.11n, I.S., Wireless LAN medium access control (MAC) and physical layer (PHY) specifications: enhancements for higher throughput, in IEEE Standard 802.11n, IEEE, 2009.

[19] DVB-S2, E.S., Digital Video Broadcasting (DVB); Second generation framing structure, channel coding and modulation systems for Broadcasting, Interactive Services, News Gathering and other broadband satellite applications (DVB-S2), in European Standard DVB-S2, European Broadcasting Union, 2009.

[20] L. Yang, et al. An FPGA implementation of low-density parity check code decoder with multi-rate capability. in conference on Asia South Pacific design automation, Shanghai, China: ACM, 2005.

[21] Y. Lei, L. Hui, C.J.R. Shi, Code construction and FPGA implementation of a low-error-floor multi-rate low-density Parity check code decoder, IEEE Trans. Circuits Syst. I Regul. Pap. 53 (4) (2006) 892–904.

[22] F. Charot, et al. A New Powerful Scalable Generic Multi-Standard LDPC Decoder Architecture. in 16th International Symposium on Field-Programmable Custom Computing Machines. Palo Alto, CA, 2008.

[23] V.A. Chandrasetty, S.M. Aziz. A reduced complexity message passing algorithm with improved performance for LDPC decoding. In: 12th International Conference on Computers and Information Technology. Dhaka, 2009.

LDPC decoding algorithms

4

4.1 STANDARD DECODING ALGORITHMS

The decoding of an LDPC code involves passing of messages between the nodes along the edges in the Tanner graph. This class of decoding algorithms is often called *message passing algorithm*. Each of the nodes in the Tanner graph works in isolation with information available along the connected edges only. These decoding algorithms require passing of the messages between the nodes to and fro for a fixed number of times or till the result is achieved. Hence such algorithms are also known as *iterative algorithms* [1].

It was stated in Section 1.1 (Fig. 1.1) that the receiver side of a Digital Communication System has a demodulator. Normally, the output from the digital demodulation block is in floating-point format. Using the floating-point format further in the channel decoding becomes extremely complex and leads to massive hardware requirement. Hence, as shown in Fig. 4.1, certain LDPC decoding algorithms operate by either making *Hard-decision* or *Soft-decision* on the messages received from the demodulator. In the former case, a binary hard-decision is made on the data received from the channel, before it is passed to the decoder. However, in the case of soft-decision algorithms, the messages are represented in the form of *Log-Likelihood Ratios* (LLR), which are quantized and then passed to the decoder in fixed-point format.

It is well known that the decoders using soft-decision method perform better compared than those using hard-decision, therefore the former is the preferred method for decoding algorithms [2]. The BER performance of un-coded BPSK, soft-decision and hard-decision based algorithms are shown in Fig. 4.2. This graph has been generated using a software simulation model in MATLAB environment.

4.1.1 BIT-FLIP ALGORITHM

The Bit-Flip (BF) algorithm is a hard-decision based LDPC decoding algorithm [1]. A binary hard-decision is done on the encoded message received from the channel and then passed on to the decoder. The messages exchanged between the check nodes and variable nodes are single bit. Each variable node (V) sends the bit information to the connected check nodes (C) over the Tanner graph

Resource Efficient LDPC Decoders. DOI: https://doi.org/10.1016/B978-0-12-811255-7.00004-6

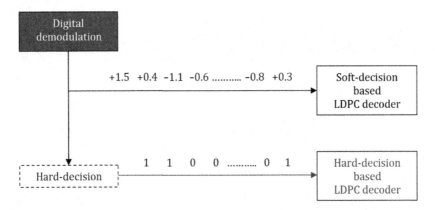

FIGURE 4.1

Hard-decision versus soft-decision based LDPC decoders.

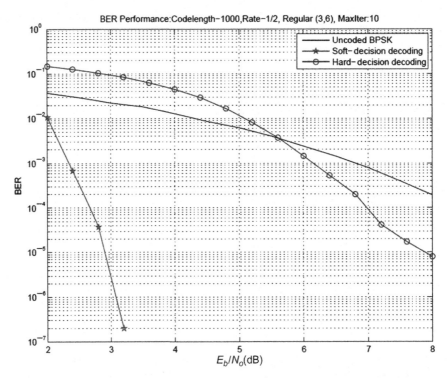

FIGURE 4.2

BER performance of hard-decision and soft-decision based LDPC decoders.

edges (refer to Fig. 2.2). Based on the bits received from the variable nodes (Eq. 4.1), each check node performs a parity check operation using modulo-2 addition (Eq. 4.2) [1]. Every check node sends back information to each of the variable nodes connected to that check node with a suggestion of the expected bit value for the parity check to be satisfied. A variable node receives a set of responses, i.e., suggested bit values from the check nodes connected to it. If the majority of the suggested bits are 1 then the variable node sets its value to 1, and if the majority of the suggested bits are 0 then the variable node has a value to 0 (Eq. 4.1) [1]. If none of the above is true then the variable node retains its original value. The above operations are repeated until the parity check is satisfied or the maximum number of iterations allowed is reached. The Bit-Flip algorithm has simple check node and variable node operations, thus making it the least complex for hardware implementation. However, the algorithm suffers from poor BER performance.

Variable node operation:

$$V_n = \begin{cases} 0 & \text{If, majority}(C_i) = 0 \\ 1 & \text{If, majority}(C_i) = 1 \\ V_n & \text{Otherwise,} \end{cases} \tag{4.1}$$

where $n = 1, 2, \ldots, N$ (variable nodes), $i = 1, 2, \ldots, d_v$ (degree of variable node "n")

Check node operation:

$$C_k = V_1 \oplus V_2 \oplus \ldots \oplus V_l \quad \forall l \neq k \tag{4.2}$$

where $l, k = 1, 2, \ldots, d_c$ (degree of check node).

4.1.2 SUM-PRODUCT ALGORITHM

The Sum-Product (SP) algorithm for LDPC decoding is a soft-decision based message-passing algorithm. The message in the decoding process is represented by the probability of each of the input bit received. The input bit probabilities are called a priori probabilities for the received bits, as the probabilities are known before initiating the LDPC decoding process. The bit probabilities returned by the decoder are called as *a posteriori* probabilities. In Sum-Product decoding, these probabilities are known as *Log-Likelihood Ratio* (LLR). The advantage of representing the probabilities in the logarithmic domain is that when probabilities need to be multiplied, the LLRs just need to be added, hence reducing the computational complexity. The aim of the SPA is to compute the maximum *a posteriori* probability for each of the input encoded bits. The information obtained by parity checks performed on the LLRs in the decoding process is known as *extrinsic message* of that particular encoded bit [1].

The above operations are collectively performed by the check nodes and variable nodes. To begin with the decoding process, the LLRs are handed over to the variable nodes (V). The variable nodes perform the operations shown in

Eq. 4.3 [3] on the input LLRs and send the results along the edges of the Tanner graph (refer to Fig. 2.2) to the connected check nodes (C) as per the parity check matrix. The check nodes perform the parity check on the messages received from the variable nodes. The check nodes also perform the operations shown in Eq. 4.4 [3] on these messages and send the information (extrinsic messages) back to the variable nodes. This process is repeated till the parity check is satisfied or a certain number of iterations is completed.

Variable node operation:

$$V_i = \text{LLR}_n + \sum_{j \neq i} C_j \tag{4.3}$$

where $n = 1, 2, \ldots, N$ (variable nodes), $i, j = 1, 2, \ldots, d_v$ (degree of variable node "n").

Check node operation:

$$C_k = 2 \tanh^{-1} \left(\prod_{k \neq l} \tanh \frac{V_l}{2} \right) \tag{4.4}$$

where $l, k = 1, 2, \ldots, d_c$ (degree of check node).

4.1.3 MIN-SUM ALGORITHM

In a practical system, the variable node operation of the sum-product algorithm (Eq. 4.3) is simple to implement as it consists of only addition operations. However, the check node operation (Eq. 4.4) is very complex to implement, as it is difficult to realize the two nonlinear functions tanh and tanh^{-1} in hardware. The simplest way to implement these complex functions is by using look-up tables (LUT). The check node equation is simplified to obtain modified version of SPA, known as the Min-Sum (MS) algorithm. The check node operation in MS is shown in Eq. 4.5. It is much simpler than that in SPA and is suitable for hardware implementations [4].

Check node operation:

$$C_k = \prod_{l \neq k} \text{sign}(V_l) \times \min_{l \neq k} |V_l| \tag{4.5}$$

where $l, k = 1, 2, \ldots, d_c$ (degree of check node).

The BER performance loss for MSA compared to SPA is about 0.3 dB at bit errors of 10^{-4} [4]. The difference is acceptable when the implementation complexity is greatly reduced. Further modifications can be done on MSA to improve the BER performance [5].

A number of variations and modifications of the MS algorithm have been reported to date, such as Offset Min-Sum [6], Normalized Min-Sum [7], Adaptive quantization in Min-Sum [8]. These modifications are proposed to improve the BER performance of the MS algorithm. However, the primary issue of complexity of practical implementations still remains a challenge.

4.1.4 STOCHASTIC ALGORITHM

The Stochastic Decoding (SD) algorithm is based on soft-decision LDPC decoding. Stochastic decoding represents the LLR as Bernoulli sequences, which encodes a probability of 1s in a sequence of bits. For a sequence of N bits, if k bits are 1s, then the probability P of the sequence is given by $P = (k/N)$. The sequence of bits regenerated from this probability mass function is not unique, however the number of 1s in the sequence is constant. In order to use this technique, the LLR is required to be converted to the probability domain and then generate the stochastic streams. Since the stochastic stream of bits does not require framing, the bits can be passed over a single edge of the Tanner graph, thus reducing hardware routing complexity of the decoder [9]. For the stochastic decoder, operations of the check node and the variable node, each with three edges (degree 3), are given in Eqs. 4.6 and 4.7 respectively [10].

Check node operation:

$$P_c = P_1(1 - P_2) + P_2(1 - P_1) \tag{4.6}$$

Variable node operation:

$$P_v = \frac{P_1 P_2}{P_1 P_2 + (1 - P_1)(1 - P_2)} \tag{4.7}$$

where P_1 = Probability of LLR1 represented by stochastic stream, P_2 = Probability of LLR2 represented by stochastic stream, P_c = Output of the check node on the third edge of that node, P_v = Output of the variable node on the third edge of that node

A comparison of different LDPC decoding algorithms based on performance and complexity is shown in Table 4.1. Note that all the algorithms listed in the

Table 4.1 Performance and Complexity of LDPC Decoding Algorithms

	Bit-Flip	Sum-Product	Min-Sum/ Layered	Stochastic
Check node operation	XOR	tanh and tanh^{-1}	XOR and comparison	XOR
Variable node operation	Comparison	Addition	Addition	Stochastic sequence
LLR quantization	N-bit	N-bit	N-bit	N-bit
Extrinsic message	1-bit	N-bit	N-bit	1-bit
BER performance	Poor	Best	Good	Moderate
Implementation complexity	Simple	Complex	Complex for long codes	Complex for Stochastic sequence generation
Clocks per decoding iteration	1	1	1	N

table require the same LLR quantization (*N-bit*) for intrinsic messages (values used by variable node at the beginning of the decoding process). The BER performance of the algorithms that use the same quantization (*N-bit*) for extrinsic messages is better compared to the other algorithms that use shorter quantization. This clearly indicates the impact of message quantization on the performance of the decoder. It is also noted that Stochastic based decoders requires N clocks per decoding iteration compared to others that require only one clock per iteration. This is due to the bit-serial processing of extrinsic messages between the nodes in stochastic algorithms.

4.2 REDUCED COMPLEXITY ALGORITHMS

A number of algorithms with varying complexity and performance have been proposed for LDPC decoding [9,11,12]. However, achieving a balanced trade-off between decoding performance (such as BER and number of iterations) and implementation complexity still remains a potential problem [13,14]. The Bit-Flip (BF) algorithm [15], which is based on hard-decision decoding, has the least complexity, but suffers from poor performance. A number of modifications have been proposed to improve its performance [16−19]. The improved bit-flipping technique presented in [16] requires dynamic computation of probabilities for bit-flipping at the variable node. Similarly, the weighted bit-flip (WBF) based algorithms [17−19] require updating of reliability values during the decoding process. Hence these modified BF algorithms require relatively complex operations compared to original BF and can achieve only modest improvement in decoding performance. In contrast, the Sum-Product (SP) algorithm, which is based on soft-decision decoding, achieves excellent decoding performance, but with very high complexity [4]. Many modifications have been proposed to simplify the node operations in SPA. The check nodes are simplified by reducing the nonlinear function to an approximated quantization table [3,20] and even to logarithmic functions [21]. But, the reduction in implementation complexity achievable by using quantization table or logarithmic functions appears to be insignificant. The check node operation of SPA is significantly simplified in the Min-Sum (MS) algorithm [4]. Simple arithmetic and logical operations required by MSA render it suitable for hardware implementation. However, the performance of the algorithm can be significantly affected by the quantization of the soft-input messages [22]. The soft messages are often quantized with higher precision to achieve higher BER performance. However, with higher level of quantization the hardware resource requirement grows. Reducing the quantization leads to reduction in hardware resources, but this comes with degradation in decoding performance, i.e., poorer BER performance and higher average iterations. This is why alternative simplified LDPC algorithms that use low-level quantization, but can still deliver improved performance need to be rigorously studied.

The majority of LDPC hardware architectures reported to date [23−27] use high precision quantized soft-input messages that require large amounts of resources. Some of these are ASIC implementations [26,27]. In [28] dynamic scaling of intrinsic/extrinsic messages is proposed and [8] presents adaptive quantization of intrinsic messages. However, the dynamic scaling and adaptive quantization of messages introduce substantial complexities in practical implementation and hardware resource overhead of the decoder. A number of other techniques have been proposed to reduce routing congestion and inter-connect complexity in the LDPC decoders [29−31]. The pulse-width message encoding technique [29] reduces the routing congestion by exchanging extrinsic messages serially between the nodes. The split-row algorithm [30] and circular-shift network technique [31] require custom LDPC matrices to alleviate the interconnect complexity. The above provides a strong rationale for further reducing the complexity of LDPC decoding algorithms. With this in mind, the following sections present two reduced complexity algorithms along with the analysis of their performance.

4.2.1 SIMPLIFIED MESSAGE PASSING

The Sum-Product algorithm performs complex node operations to provide significantly better BER performance compared to the less complex Bit-Flip algorithm. A Simplified Message Passing (SMP) algorithm is presented in this section [32,33], which combines features of SPA and BF algorithms. Obviously, the aim of the SMP algorithm is to achieve better decoding performance than Bit-Flip at a lower implementation complexity compared to the Sum-Product algorithm. The check node and variable node operations of the SMP algorithm are presented next.

4.2.1.1 Check node operation

The complexity of a message passing algorithm is significantly affected by the quantization length of the extrinsic messages and the complexity of the check node operations [22]. These aspects are particularly critical in case of hardware implementation of large LDPC codes. In order to reduce the complexity, the check node in the SMP algorithm consists of a simple parity check operation (Eq. 4.8) [34] requiring XOR logic only, similar to the BF algorithm. The performance improvement of the SMP algorithm over BF is achieved from a distinct variable node (V) operation described later.

$$C_k = V_1 \oplus V_2 \oplus \cdots \oplus V_l \quad \forall l \neq k \tag{4.8}$$

where $l, k = 1,2,\ldots$ degree of check node.

Note that the stochastic [25] and binary message-passing [35−37] based LDPC decoders also incorporate a similar check node operation requiring simple XOR logic. However, these techniques propose using serialized messages between the variable and check nodes, and therefore require extra hardware (e.g., FIFO) and additional clock cycles, leading to a very slow decoding process. In comparison, the SMP algorithm uses only single-bit messages where serialization is not required.

4.2.1.2 Variable node operation

As stated previously, fully hard-decision based decoding algorithm such as the Bit-Flip suffers from poor BER performance because the intrinsic and extrinsic messages used in the decoding process are based on hard-decisions. In the SMP algorithm, performance improvement compared to fully hard-decision based algorithms [2] is achieved by using soft log-likelihood ratios (LLR) as primary inputs to the decoders, like any other soft-decision based algorithms [3]. The variable node (V) performs "sum" operation similar to the Sum-Product algorithm, but the difference is that the SMP algorithm requires the original LLR value only at the beginning of the decoding cycle, as in Eq. 4.9. In subsequent iterations, new (updated) LLR values calculated by Eq. 4.12 are used. As described in the next paragraph, the LLR values are updated by simply adding or subtracting a constant integer in contrast with SPA where the variable node operation utilizes two non-linear functions tanh and $\tanh^{-1}$ computed by the check nodes.

In the analysis of hard-decision based algorithms presented in [38], it is apparent that the variable node operates directly on the hard-decision bit received from the check nodes. In contrast, the SMP algorithm presented here maps the bit suggestion from the check nodes (C) to an optimized integer constant called "Weight" ($\pm W$). The value of W is determined from simulations to achieve the best possible BER performance. It is either added to or subtracted from the current LLR value. For example, in a variable node, if binary "0" is received from the check node, it is mapped to $+W$, whereas, if a binary "1" is received, it is mapped to $-W$, as shown in Eqs. 4.10 and 4.11 respectively. After this operation, a hard-decision is performed on the updated LLR value and the resulting 1-bit message is sent across to the respective check nodes for parity check, as shown in Eq. 4.13. The process is repeated until the parity check is satisfied or the maximum iteration is reached.

Initial (at iteration "0"):

$$V_n = LLR \tag{4.9}$$

Subsequent iterations:

$$X_i = +W, \quad \text{If} \quad C_i = 0 \tag{4.10}$$

$$X_i = -W, \quad \text{If} \quad C_i = 1 \tag{4.11}$$

$$V_n = V_n + \sum X_i \tag{4.12}$$

$$V_i = \text{sign}(V_n - X_i) \tag{4.13}$$

$$[\,''+'' \rightarrow ''0'' \text{ and } ''-'' \rightarrow ''1''\,]$$

where $n = 1,2,\ldots$ number of variable nodes, $i = 1,2,\ldots$, degree of variable node, W = optimized integer constant obtained from simulations.

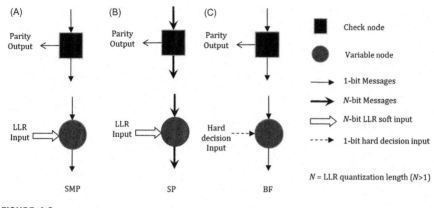

FIGURE 4.3

Comparison of decoding node structures for (A) SMP, (B) SP, and (C) BF algorithms.

Clearly, in the SMP algorithm, the check node performs much simpler XOR operation compared to complex computations of tanh and $\tanh^{-1}$ functions in SPA. Also, in SMP, the variable node performs addition operations and uses simple mapping logic compared to SPA where the variable node relies on complex check node computations. Hence the SMP algorithm can be implemented using simple hardware blocks, such as adders and Look-Up-Tables (LUT).

A comparison of check node and variable node structures for the SMP algorithm with those of the Sum-Product and Bit Flip algorithms is shown in Fig. 4.3.

4.2.2 MODIFIED MIN-SUM

Although the simplified check node operation in Min-Sum (MS) algorithm has reduced the complexity compared to the Sum-Product (SP) algorithm, the former still exchanges high precision messages between the decoding nodes in order to achieve decoding performance comparable to that of SP. The level of quantization used in the LLR and extrinsic messages of MS directly impacts upon the decoding performance. As the quantization length of the message decreases, both the BER performance and complexity of the algorithm reduces [32]. Studies have shown that there is slight performance loss in going from 5-bit to 4-bit or even to 3-bit quantization [22]. Using 2-bit quantization leads to massive reduction in implementation complexity, but at a significant loss in decoder performance compared to 3-bit quantization. The performance of 2-bit MS has been improved through extrinsic message optimization reported in [11]. The performance is further improved by the Modified Min-Sum (MMS) algorithm presented in the remainder of this section [39,40]. The MMS uses higher precision LLR messages while exchanging lower precision (2-bit) extrinsic messages. The check node and variable node operations of the MMS algorithm are described next.

4.2.2.1 Variable node operation

The variable node operation in MMS is similar to that in the original Min-Sum algorithm [4]. The modification introduced in the MMS algorithm is that the variable node (V) performs operations on LLRs that have high precision (high quantization length), but maps the computed result to a 2-bit message to be passed on to the check nodes, as shown in Eq. 4.14. The 2-bit message consists of a sign bit and a magnitude bit representing the computed *LLR sum*. The mapping is shown in Eq. 4.15 and is based on a threshold (T_m) obtained from simulations. Depending on the message received from the check nodes (C_j), the 2-bit information is again mapped to constant values ($\pm W$ or $\pm w$) to perform the *LLR sum* operation in the variable node. This mapping is shown in Eq. 4.16. The constant values for mapping are also obtained from simulations.

$$V_i = g\left(\text{LLR}_n + \sum_{j \neq i} f(C_j)\right) \tag{4.14}$$

where $n = 1, 2, \ldots, N$ (variable nodes), $i, j = 1, 2, \ldots, d_v$ (degree of variable node "n")

$$g(y) = \begin{cases} 01 & \text{if,} & y > T_m \\ 00 & \text{if,} & 0 \leq y \leq T_m \\ 10 & \text{if,} & 0 > x \geq -T_m \\ 11 & \text{if} & x < -T_m \end{cases} \tag{4.15}$$

$$f(x) = \begin{cases} +W & \text{if,} & x = 01 \\ +w & \text{if,} & x = 00 \\ -w & \text{if,} & x = 10 \\ -W & \text{if,} & x = 11 \end{cases} \tag{4.16}$$

where T_m is the optimized threshold for mapping obtained from simulations; W is the optimized higher integer constant obtained from simulations; w is the optimized lower integer constant obtained from simulations.

4.2.2.2 Check node operation

In the conventional MS algorithm, the check node is expected to determine the product of the sign of incoming messages and also find the minimum of the magnitudes of the input messages [4]. In the MMS algorithm presented here, the product of the signs of the incoming messages is computed using XOR operation (S) and the minimum magnitude is determined using AND operation (M), as shown in Eqs. 4.17 and 4.18 respectively. The check node output message (C) is obtained simply by concatenating the sign bit and the magnitude bit, as shown in Eq. 4.19. The message passing between the nodes continues till the parity check is satisfied or maximum iteration is reached. In the check node operation of conventional MS algorithm, the quantization used for extrinsic messages is the same as that used for intrinsic messages. Whereas, the MMS algorithm presented here uses reduced quantization (2-bit) for extrinsic messages compared to that used for intrinsic messages.

$$S_k = V_1^{(s)} \oplus V_2^{(s)} \oplus \cdots \oplus V_l^{(s)} \quad \forall l \neq k \tag{4.17}$$

$$M_k = V_1^{(m)} \& V_2^{(m)} \& \cdots \& V_l^{(m)} \quad \forall l \neq k \tag{4.18}$$

$$C_k = \{S_k M_k\} \tag{4.19}$$

where $l, k = 1, 2, \ldots, d_c$ (degree of check node); $S =$ Sign bit of check node message; $M =$ Magnitude bit of check node message; $V_l^{(s)} =$ Sign bit of the message "l" from variable node; $V_l^{(m)} =$ Magnitude bit of the message 'l' from variable node.

The message mapping in the variable node described above is similar to that presented in [11,41]. However, the differences in the MMS algorithm presented here are that:

1. It uses higher precision intrinsic messages (LLR) to improve the decoding performance.
2. It requires higher precision addition operation (instead of 2-bit) in the variable node due to the use of higher precision intrinsic messages.
3. It simplifies the variable node operation by eliminating the scaling factor.
4. It incorporates simple "AND" logic instead of comparators to determine the minimum of the magnitudes of the input messages at the check node.

All these modifications lead to simplified check node and reduced inter-connect complexity. Since the check node operation in Eq. 4.18 requires comparing single-bit inputs to determine the minimum, a simple "AND" operation is sufficient to perform the comparison. Compared to conventional Min-Sum decoders that use 2-bit quantization for both intrinsic and extrinsic messages, the MMS algorithm has slightly increased complexity in the variable node due to the use of higher precision LLRs (intrinsic messages). However, the impact of this increased complexity is insignificant when compared to the advantages gained in the improvement of overall performance and complexity over conventional Min-Sum decoders.

A comparison of check node and variable node structures for the MMS algorithm presented here with those of existing MS and other 2-bit based algorithms is shown in Fig. 4.4.

4.3 PERFORMANCE ANALYSIS OF SIMPLIFIED ALGORITHMS

To ascertain the performance improvements resulting from the Simplified Message Passing (SMP) and Modified Min-Sum (MMS) algorithms presented in the previous section compared to the other benchmarked algorithms, software simulation models have been developed using the C programming language in the

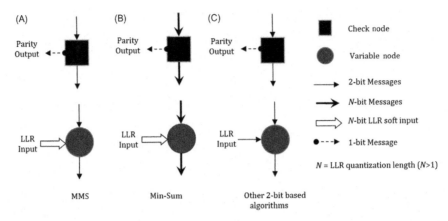

FIGURE 4.4

Comparison of decoding node structures for (A) MMS, (B) MS, and (C) other 2-bit based algorithms.

MATLAB environment. The models represent ½ rate (3, 6) regular LDPC decoders [42] with parameterized code length and are bit-true simulation models to handle quantization of the messages. The LDPC codes used in the simulations were generated using Progressive Edge Growth (PEG) algorithm [43]. Simulations were carried out assuming that the codewords were modulated using Binary Phase Shift Keying (BPSK) and passed over an Additive White Gaussian Noise (AWGN) channel [44]. The maximum number of iterations allowed for decoding was set to 10. Each of the BER data points with respect to E_b/N_o is obtained by running the simulations until at least 1000 erroneous bits are identified.

4.3.1 EXTRACTION OF OPTIMIZED PARAMETERS

Simplified Message Passing: As stated previously, it is very important to use the optimum value of "W" in order to achieve the best possible BER performance in a decoder based on the SMP algorithm (see Eqs. 4.3 and 4.4). Monte Carlo simulations were carried out using different LLR quantization (3-bit, 4-bit and 5-bit) with various values of W (1 to 5) and for different E_b/N_o levels. Fig. 4.5 shows the simulation results. A ½ rate (3, 6) 648-bit LDPC code with a maximum iteration of 10 was used in the simulations. From Fig. 4.5, it is clear that SMP with 3-bit LLR quantization can achieve the lowest BER at $W = 1$, whereas SMP with 4-bit LLR quantization has optimum BER performance at both $W = 1$ and $W = 2$. The BER performance of SMP with 5-bit LLR quantization is almost constant over a wide range of "W", achieving the best performance at $W = 2$. It is also observed that SMP has similar performance at $W = 1$, irrespective of the LLR quantization length.

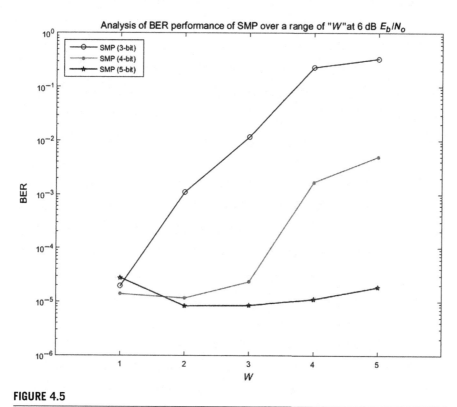

FIGURE 4.5

BER performance of SMP algorithm over a range of "W".

Modified Min-Sum algorithm: Monte Carlo simulations were carried out on a decoder based on the Modified Min-Sum (MMS) algorithm presented in the previous section. The simulations used 4-bit LLRs and were carried out to ascertain the impact of T_m, W and w values on BER performance. The simulation results at $E_b/N_o = 4$ dB are shown in Fig. 4.6. Some of the combinations of values of T_m, W and w returning relatively better BER performances are summarized in Table 4.2.

The BER and average iteration performance of the MMS decoder for the majority of the configurations in Table 4.2 (C1, C2, C3, and C4) are shown in Figs. 4.7 and 4.8 respectively. A ½ rate 1000-bit LDPC code with a maximum of 10 iterations has been used in the simulations. From Fig. 4.7, it is clear that the MMS decoder with configuration C1 has optimum BER performance compared to the other configurations shown in the figure. Similarly, from Fig. 4.8, it is noted that the MMS decoder with configuration C3 requires the least number of decoding iterations at 4 dB E_b/N_o.

Note that the MMS algorithm presented above is based on simplification of extrinsic messages, which leads to modification of variable/check node operations irrespective of the modulation scheme used in the communication system. The

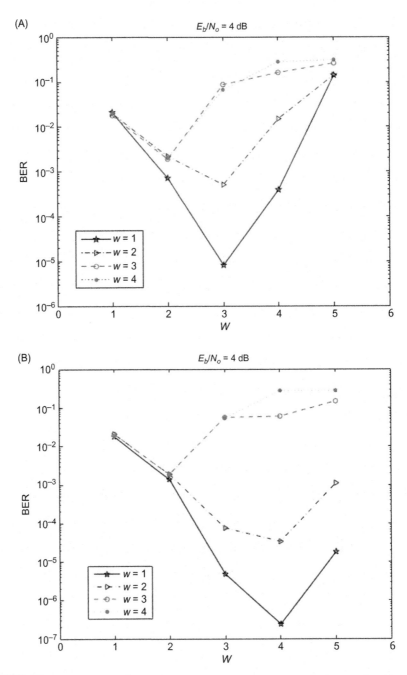

FIGURE 4.6

BER performance for various values of T_m, W and w. (A) Effect of W and w values on BER performance at $T_m = 1$. (B) Effect of W and w values on BER performance at $T_m = 2$. (C) Effect of W and w values on BER performance at $T_m = 3$.

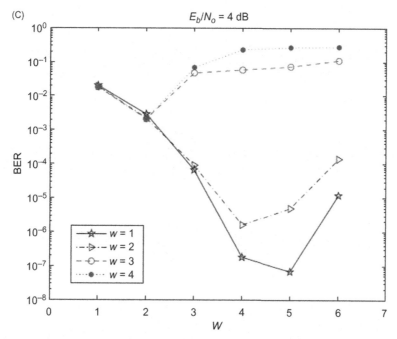

FIGURE 4.6

(Continued)

Table 4.2 Parameter Configurations for MMS Decoder Delivering Relatively Higher BER Performance

Configurations	BER	T_m	W
C1	6.8×10^{-8}	3	5
C2	1.8×10^{-7}	3	4
C3	2.4×10^{-7}	2	4
C4	4.8×10^{-6}	2	3
C5	8.2×10^{-6}	1	3

$w = 1$ and $E_b/N_o = 4$ dB.

effect on algorithmic performance due to change in modulation scheme (e.g., QPSK, 16QAM) will have little or no impact on the design parameters (T_m, W and w). The parameters are expected to vary based on the quantization of intrinsic and extrinsic messages. So the technique presented here for optimization of T_m, W and w should be applicable to other modulation schemes.

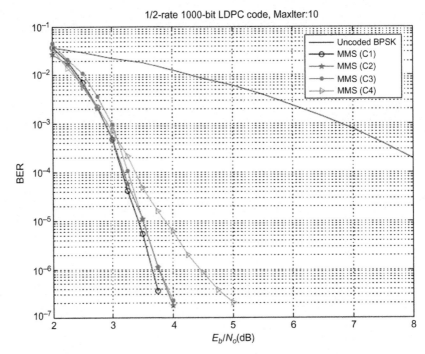

FIGURE 4.7

BER performance of MMS algorithm with various configurations.

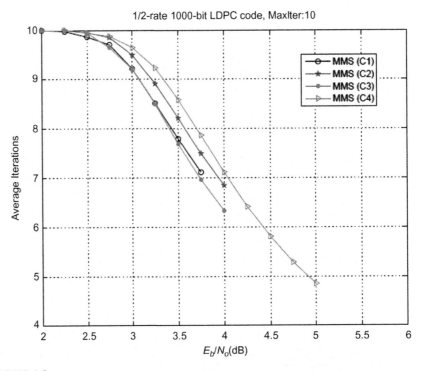

FIGURE 4.8

Average iterations for MMS algorithm with various configurations.

4.3.2 **PERFORMANCE COMPARISON**

Simplified Message Passing: Simulation results on BER and frame error rate (FER) performance for the SMP algorithm are shown in Figs. 4.9 and 4.10 respectively. A 1000-bit LDPC code with LLR quantization of 3-bit, 4-bit (Weight, $W = 1$) and 5-bit (Weight, $W = 2$) have been used in the simulations. Key features of these results are summarized in Table 4.3. Clearly, the SMP algorithm achieves better BER performance compared to Bit-Flip (BF). At a BER of 10^{-5}, the improvement in SMP using 3-bit LLR is 1.1 dB E_b/N_o (Fig. 4.9). Higher LLR precisions (4-bit and 5-bit) in the variable node operations improve the BER performance by at least 2 dB compared to BF at a BER of 10^{-5} (Fig. 4.9). The SMP algorithm improves the frame error rate performance over BF in a similar fashion as observed from Fig. 4.10 and Table 4.3. The convergence rate of various algorithms have been assessed by analyzing the average number of decoding iterations required by each algorithm, as shown in Fig. 4.11. Clearly, the SMP algorithm with 4-bit and 5-bit LLR precisions require much fewer iterations (higher convergence rate) compared to BF, for example at 6 dB E_b/N_o. However, with 3-bit LLR precision, SMP requires more iterations

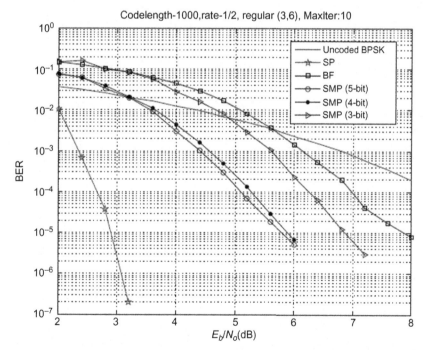

FIGURE 4.9

BER performance for the SMP, SP and BF algorithms.

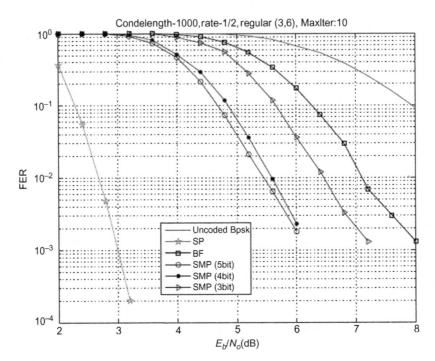

FIGURE 4.10

FER performance for the SMP, SP and BF algorithms.

Table 4.3 Performance Comparison of SMP and Other Algorithms

Algorithms		BER of 10^{-5} (dB)	FER of 10^{-2} (dB)	Avg. Iterations at BER of 10^{-5}
SP		2.9	2.6	6.8
BF		7.9	7.1	2.4
SMP	3-bit	6.8	6.5	4.5
	4-bit	5.9	5.6	4.3
	5-bit	5.8	5.5	4.2

compared to BF for E_b/N_o greater than 6 dB. Although the iteration count for Sum-Product (SP) algorithm is much lower than SMP, each of the iterations in SP is likely to take significantly more computation time due to the highly complex operations at the variable and check nodes (see Eqs. 4.3 and 4.4). In contrast, the SMP algorithm has much simpler node operations (see Eqs. 4.8−4.12) and therefore incurs much shorter iteration cycle time.

Modified Min-Sum: The BER performance of the MMS algorithm (for configurations C1, C2, C3 and C4) compared to the original Min-Sum (MS) and other

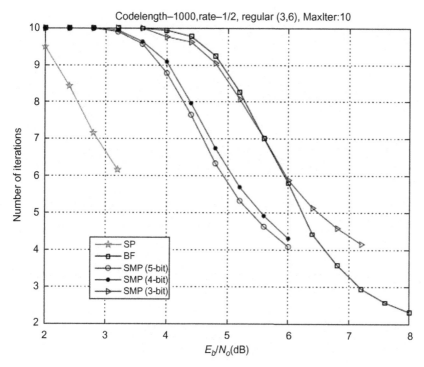

FIGURE 4.11

Average decoding iterations for SMP, SP and BF algorithms.

decoders [11] (see Ref. A and Ref. B) are shown in Fig. 4.12. A 1000-bit LDPC code with a maximum of 10 iterations has been used in the simulation of MMS (4-bit LLR quantization) and the MS decoders. The simulation results are summarized in Table 4.4 to aid the clarity of the analysis presented here. At a BER of 10^{-6}, MMS (C1) achieves a gain of 2.5 dB over MS (2-bit) and about 0.2 dB over both MS (3-bit) and optimized 2-bit. However, it suffers a loss of 0.2 dB compared to MS (4-bit) and 0.5 dB over Sum-Product. FER performance improvement is also observed in MMS (C1) compared to MS (2-bit and 3-bit) based decoders, as shown in Fig. 4.13. The average decoding iterations required by these algorithms are shown in Fig. 4.14. A significant improvement of average decoding iterations for MMS (C1) compared to MS (2-bit) can be observed. MMS (C1) has average iterations close to that of MS (3-bit).

Note that MS (4-bit) has better decoding performance (BER and iterations) compared to MMS (C1). Although, MMS (C1) and MS (4-bit) use 4-bit intrinsic messages, MMS (C1) uses only 2-bit extrinsic messages for decoding operation. This technique proposed in the MMS significantly reduces the implementation complexity and also saves hardware resources.

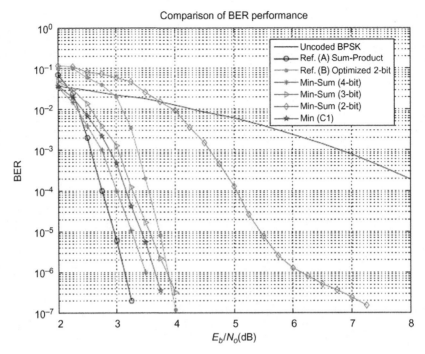

FIGURE 4.12

Comparison of BER performance of MMS with other decoders.

Table 4.4 Performance Comparison of MMS and Other Algorithms

Algorithms		LDPC Code	Maximum Iterations	E_b/N_o (dB) at BER of 10^{-6}	Avg. Iterations at BER of 10^{-6}
MMS (4-bit)	C1	½ rate 1000-bit	10	3.65	7.6
	C2			3.75	7.5
	C3			3.75	7.0
	C4			4.45	5.8
Min-Sum	4-bit			3.50	6.6
	3-bit			3.80	6.9
	2-bit			6.10	5.3
Ref. B [11] Optimized 2-bit		½ rate 1974-bit	120	3.90	9.0
Ref. A [11] Sum-Product		½ rate 1974-bit	120	3.10	5.0

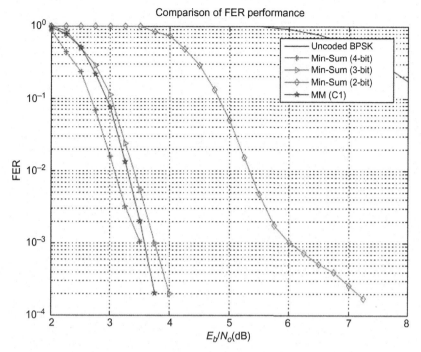

FIGURE 4.13

Comparison of FER performance of MMS with other decoders.

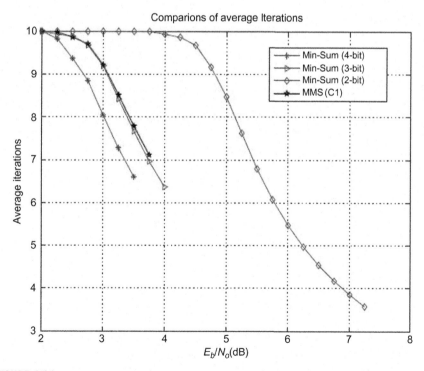

FIGURE 4.14

Comparison of average decoding iterations of MMS with other decoders.

4.4 SUMMARY

A Simplified Message Passing (SMP) algorithm for LDPC decoding has been presented in this chapter. This algorithm uses higher precision soft LLR-inputs for variable node operations while passing only hard-decision messages between the processing nodes. The algorithm has been verified through software simulations. The results show that the SMP algorithm leads to improved BER and FER performances compared to fully hard-decision based solutions such as those based on the Bit-Flip algorithm. The SMP algorithm also reduces the average number of decoding iterations compared to Bit-Flip. SMP incorporates highly simplified node operations with single-bit extrinsic messages. It therefore results in significantly reduced complexity in hardware implementations of the decoder.

This chapter also presented an innovative way to reduce the extrinsic message length in LDPC decoders without incurring the level of BER degradation normally associated with reduction in intrinsic message quantization. It has been shown that the proposed Modified Min-Sum (MMS) decoder with 2-bit extrinsic messages and 4-bit intrinsic messages achieves BER performance slightly better than that of a 3-bit Min-Sum decoder. The software simulation results validate these claims. The results also demonstrate that the MMS decoder addresses the problem of massive degradation in BER performance resulting from a decoder based on straightforward realization of 2-bit Min-Sum algorithm where both intrinsic and extrinsic messages are 2-bit long. Maintaining a higher quantization at the intrinsic message helps to provide a higher BER performance. The MMS scheme can be adapted to appropriate combinations of intrinsic and extrinsic message quantization to suit particular applications. If desired, the BER performance can be improved further by using a different combination of intrinsic and extrinsic message quantization. This of course may require a larger amount of hardware resources, but the resource requirement will still be much less than that of a straightforward implementation of Min-Sum.

After developing the reduced complexity decoding algorithms, the next logical step would normally be to use them to develop resource efficient hardware architectures. However, the complexity of hardware implementation and flexibility of the decoder greatly depend on the structure of the LDPC matrix. Therefore, the next chapter presents LDPC matrix construction techniques to aid in the design of resource efficient LDPC decoders, particularly focusing on partially-parallel hardware architectures.

REFERENCES

[1] S.J. Johnson, Introducing Low-Density Parity-Check Codes, University of Newcastle, Australia, 2006.
[2] M. Singh, I.J. Wassell. Comparison between soft and hard decision decoding using quaternary convolutional encoders and the decomposed CPM model. in IEEE VTS 53rd Vehicular Technology Conference, Rhodes, 2001.

[3] S. Papaharalabos, et al., Modified sum-product algorithms for decoding low-density parity-check codes, IET Commun. 1 (3) (2007) 294−300.

[4] A. Anastasopoulos. A comparison between the sum-product and the min-sum iterative detection algorithms based on density evolution. In: IEEE Global Telecommunications Conference, San Antonio, TX, 2001.

[5] Z. Zhou, et al. Modified min-sum decoding algorithm for LDPC codes based on classified correction. in: 3rd International Conference on Communications and Networking, Hangzhou, 2008.

[6] X. Meng, W. Jianhui, Z. Meng. A modified Offset Min-Sum decoding algorithm for LDPC codes. in: 3rd IEEE International Conference on Computer Science and Information Technology, Chengdu, 2010.

[7] W. Xiaofu, et al., Adaptive-normalized/offset min-sum algorithm, IEEE Commun. Lett. 14 (7) (2010) 667−669.

[8] S. Kim, G.E. Sobelman, H. Lee. Adaptive quantization in min-sum based irregular LDPC decoder. in: IEEE International Symposium on Circuits and Systems, 2008.

[9] C. Winstead, et al. Stochastic iterative decoders. in: IEEE International Symposium on Information Theory, 2005.

[10] A. Rapley, et al. Stochastic iterative decoding on factor graphs. in: 3rd International Symposium on Turbo Codes and Related Topics, 2003.

[11] Z. Cui, Z. Wang, Improved low-complexity low-density parity-check decoding, IET Commun. 2 (8) (2008) 1061−1068.

[12] G. Lechner, I. Land, L. Rasmussen. Decoding of LDPC codes with binary vector messages and scalable complexity. in: International Symposium on Turbo Codes and Related Topics, Lausanne, 2008.

[13] E. Yeo, V. Anantharam, Capacity approaching codes, iterative decoding architectures, and their applications, IEEE Commun. Mag. 41 (8) (2003) 132−140.

[14] S.M. Aziz, M.D. Pham, Implementation of low density parity check decoders using a new high level design methodology, J. Comput. Acad. Publ. 5 (1) (2010) 81−90.

[15] D.J.C. MacKay, Good error-correcting codes based on very sparse matrices, IEEE Trans. Inf. Theory 45 (2) (1999) 399−431.

[16] N. Miladinovic, M.P.C. Fossorier, Improved bit-flipping decoding of low-density parity-check codes, IEEE Trans. Inf. Theory 51 (4) (2005) 1594−1606.

[17] Q. Dajun, et al. A modification to weighted bit-flipping decoding algorithm for LDPC codes based on reliability adjustment. in: IEEE International Conference on Communications, Beijing, 2008.

[18] F. Guo, L. Hanzo, Reliability ratio based weighted bit-flipping decoding for low-density parity-check codes, Electronics Lett. 40 (21) (2004) 1356−1358.

[19] C.H. Lee, W. Wolf, Implementation-efficient reliability ratio based weighted bit-flipping decoding for LDPC codes, Electronics Lett. 41 (13) (2005) 755−757.

[20] J.H. Han, M.H. Sunwoo, Simplified sum-product algorithm using piecewise linear function approximation for low complexity LDPC decoding, 3rd International Conference on Ubiquitous Information Management and Communication, ACM, Suwon, Korea, 2009.

[21] S. Papaharalabos, P.T. Mathiopoulos, Simplified sum-product algorithm for decoding LDPC codes with optimal performance, Electronics Lett. 45 (2) (2009) 116−117.

[22] R. Zarubica, et al. Efficient quantization schemes for LDPC decoders. in: IEEE Military Communications Conference, San Diego, CA, 2008.

[23] H. Zhiyong, R. Sebastien, F. Paul. FPGA implementation of LDPC decoders based on joint row-column decoding algorithm. in: IEEE International Symposium on Circuits and Systems, New Orleans, LA, 2007.

[24] S.S. Tehrani, S. Mannor, W.J. Gross. An area-efficient FPGA-based architecture for fully-parallel stochastic LDPC decoding. in: IEEE Workshop on Signal Processing Systems, Shanghai, China, 2007.

[25] S. Sharifi Tehrani, S. Mannor, W.J. Gross, Fully parallel stochastic LDPC decoders, IEEE Trans. Signal Process. 56 (11) (2008) 5692−5703.

[26] G. Masera, F. Quaglio, F. Vacca, Implementation of a flexible LDPC decoder, IEEE Trans. Circuits Syst. II Express Briefs 54 (6) (2007) 542−546.

[27] G. Masera, F. Quaglio, F. Vacca, Finite precision implementation of LDPC decoders, IEEE Proc. Commun. 152 (6) (2005) 1098−1102.

[28] Z. Xiaojun, et al. Uniform all-integer quantization for irregular LDPC decoder. in: 5th International Conference on Wireless Communications, Networking and Mobile Computing, 2009.

[29] K. Cushon, et al., A min-sum iterative decoder based on pulsewidth message encoding, IEEE Trans. Circuits Syst. II Express Briefs 57 (11) (2010) 893−897.

[30] T. Mohsenin, D.N. Truong, B.M. Baas, A low-complexity message-passing algorithm for reduced routing congestion in LDPC decoders, IEEE Trans. Circuits Syst. I Regul. Pap. 57 (5) (2010) 1048−1061.

[31] C. Xiaoheng, L. Shu, V. Akella, QSN:a simple circular-shift network for reconfigurable quasi-cyclic LDPC decoders, IEEE Trans. Circuits Syst. II Express Briefs 57 (10) (2010) 782−786.

[32] V.A. Chandrasetty, S.M. Aziz. A reduced complexity message passing algorithm with improved performance for LDPC decoding. in: 12th International Conference on Computers and Information Technology, Dhaka, 2009.

[33] V.A. Chandrasetty, S.M. Aziz, FPGA implementation of a LDPC decoder using a reduced complexity message passing algorithm, J. Netw. Acad. Publ. 6 (1) (2011) 36−45.

[34] S. Hirst, B. Honary, Decoding of generalised low-density parity-check codes using weighted bit-flip voting, IEEE Proc. Commun. 149 (1) (2002) 1−5.

[35] N. Mobini, A.H. Banihashemi, S. Hemati. A differential binary message-passing LDPC decoder. in: IEEE Global Telecommunications Conference, Washington, DC, 2007.

[36] C. Chao-Yu, et al. A binary message-passing decoding algorithm for LDPC codes. in 47th Annual Allerton Conference on Communication, Control, and Computing, Monticello, IL, 2009.

[37] H. Qin, et al., Two reliability-based iterative majority-logic decoding algorithms for LDPC codes, IEEE Trans. Commun. 57 (12) (2009) 3597−3606.

[38] G. Lechner, T. Pedersen, G. Kramer. EXIT chart analysis of binary message-passing decoders. in: IEEE International Symposium on Information Theory, Nice, 2007.

[39] V.A. Chandrasetty, S.M. Aziz. FPGA implementation of high performance LDPC decoder using modified 2-bit min-sum algorithm. in: 2nd International Conference on Computer Research and Development, Kuala Lumpur, 2010.

[40] V.A. Chandrasetty, S.M. Aziz, An area efficient LDPC decoder using a reduced complexity min-sum algorithm, Integr. VLSI J. 45 (2) (2012) 141−148.

[41] S.K. Chilappagari, et al. Two-bit message passing decoders for LDPC codes over the binary symmetric channel. in: IEEE International Symposium on Information Theory, Seoul, 2009.

[42] R. Zarubica, S.G. Wilson, E. Hall. Multi-Gbps FPGA-based low density parity check (LDPC) decoder design. in: IEEE Global Telecommunications Conference, Washington, DC, 2007.

[43] X.Y. Hu. Software to Construct PEG LDPC code, 2008 [May 2009]. Available from: <http://www.inference.phy.cam.ac.uk/mackay/PEG_ECC.html>.

[44] J.G. Proakis, in: M. Salehi (Ed.), Digital Communications., fifth ed., McGraw-Hill, New York, 2008.

LDPC decoder architectures

5

5.1 COMMON HARDWARE ARCHITECTURES

5.1.1 FULLY-PARALLEL

In a fully parallel architecture of LDPC decoder [1], all the check node and variable node processing elements are independently instantiated as per the graph representation of the LDPC code [2]. The interconnections of the nodes are done according to the parity check matrix or the graph. Hence, each check node and variable node processor can perform their operations independently. The advantage of this kind of architecture is that decoding process is faster; hence, very high throughput can be achieved. A LDPC decoder with fully-parallel architecture has been implemented on a field programmable gate array (FPGA) using Min-Sum algorithm for ½ rate 504-bit code length achieving a throughput of more than 1 Gbps [3]. However, this kind of implementation consumes massive hardware resources, especially with Min-Sum algorithm and leads to routing complexity. In the above design, 75% of the Xilinx Virtex 2 (XC2VP100) FPGA logic resources is utilized, which is extremely high. A simple block diagram of a parallel LDPC decoder architecture is shown in Fig. 5.1. Implementation of parallel LDPC decoder architecture involves a simple and straightforward approach without much design complexity.

5.1.2 FULLY-SERIAL

A serial architecture for LDPC decoder has a totally different approach. This architecture is more conservative on hardware resource utilization by sacrificing throughput of the decoder significantly [4]. A decoder based on serial architecture requires memory for its operations. The memory access by check node and variable node processors is according to the parity check matrix. The check node and variable node processors consist of a single node each. When the LLR is loaded to the decoder, the variable node processor enables its operation by processing the data sequentially node by node and writes the output to a message memory. Once the variable node processing is completed, the check node processor is enabled by the control unit. The check node processor reads the information from the message memory for processing sequentially node by node and writes back the output to the same memory. When the check node processing is finished, the control unit enables

Resource Efficient LDPC Decoders. DOI: https://doi.org/10.1016/B978-0-12-811255-7.00005-8

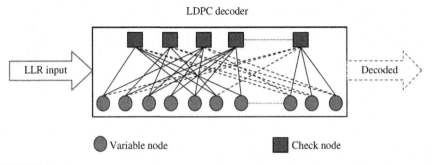

FIGURE 5.1

Block diagram of an LDPC decoder using fully-parallel architecture.

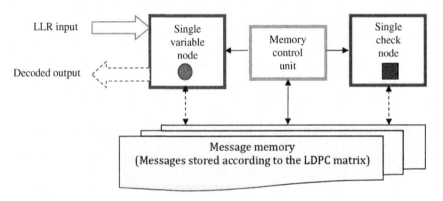

FIGURE 5.2

Block diagram of an LDPC decoder using serial architecture.

the variable node processor again for further processing. The process continues till the maximum iteration is reached or the information is decoded. A simple block diagram of a serial LDPC decoder architecture is shown in Fig. 5.2.

The serial LDPC decoder architecture uses memory to store all the intermediate results generated by each of the node processing units. An LUT, processor or a microcontroller can be used to implement the memory control unit for generation of message-memory address. In [5], a serial architecture of LDPC decoder using Min-Sum algorithm is implemented using an embedded Power PC in Virtex II FPGA.

5.1.3 PARTIALLY-PARALLEL

The partially-parallel architecture is an approach to balance the trade-off between resource utilization and throughput [6]. This architecture makes use of FPGA logic and memory resources as well. In this kind of architecture [7], only a few

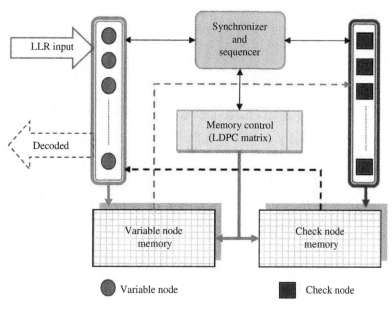

FIGURE 5.3

Block diagram of an LDPC decoder using partially-parallel architecture.

check nodes and variable nodes (also called as *decoding units*) are instantiated rather than only one node (in case of serial architecture) or all the nodes (in case of fully-parallel architecture), thus utilizing fewer hardware resource. A larger number of processing clock cycles are required by the decoder to complete its operation, ultimately reducing the overall throughput of the decoder. The intermediate results are stored in memory blocks. There can be memory access conflicts by the decoding units, due to the complex memory addressing system required for this architecture. In simple terms, the performance and design complexity of partially-parallel LDPC decoder architecture is in between the serial and fully-parallel architecture. A partially-parallel LDPC decoder is implemented on FPGA using Layered Decoding algorithm in [8] for ½ rate 1536-bit code length achieving a throughput of about 38 Mbps. A simple block diagram of partially-parallel LDPC decoder architecture is shown in Fig. 5.3.

In partially-parallel architecture, only few nodes of the LDPC decoder are designed to operate in parallel at any instance and then re-use the same structure to perform other node operations. The intermediate results generated by the node operations are required to be stored in memory, thus requiring a memory controller similar to the serial architecture. Since the set of nodes that operate in parallel are reused during the decoding process, the parallel interconnections of that set of nodes directly depend on the structure of the LDPC matrix. Hence, a structured LDPC matrix is required to be constructed in order to exploit the reusable nature of the partially-parallel architecture. Different variations of Quasi Cyclic (QC) LDPC matrices are generally used for designing a partially-parallel LDPC decoder architecture [9].

5.2 REVIEW OF PRACTICAL LDPC DECODERS

In this section, a summary of different LDPC decoder implementations on FPGA is presented based on the review of relevant publications.

In [10], a fully-parallel FPGA based LDPC decoder using Min-Sum algorithm has been designed. The design uses 1200 bit—½ rate (3, 6) code. The architecture incorporates 3-stage pipeline design having input, decoding and output with a maximum of 10 iterations and 3-bit message data. The design can operate at 6 Gbps with a clock frequency of 100 MHz and BER performance of 10^{-4} at E_b/N_o of 3.7 dB. However, the design which is synthesized for Virtex4 FPGA utilizes massive hardware resources with 45% (40,613) of slices and 10% (18,945) of flip-flops, which is a major disadvantage of fully-parallel architecture using Min-Sum algorithm.

A fully parallel FPGA based stochastic LDPC decoder is proposed in [11]. Sum-Product algorithm is used in this work. A ½ rate 16-bit irregular LDPC code is used in this decoder design. The random bits for the stochastic conversion process are generated using LFSR. A maximum throughput of 8.4 Mbps is achieved at a maximum clock frequency of 165 MHz on Altera Stratix II FPGA. The BER performance of the implemented design is about 10^{-4} at 7 dB SNR. A throughput of about 8 Mbps is too low for practical applications.

A parallel architecture for decoding 1024-bit LDPC codes is designed in [12]. Min-Sum algorithm is used in the design. The proposed architecture consists of concatenated blocks of each variable and check node processing logic units, which corresponds to a single iteration, making the system highly parallel to speed up the operations. The design is synthesized for Virtex5 FPGA, leading to a resource utilization of 6.95% (14,402) LUTs and 8.18% (16,952) registers. The authors have not made performance analysis on the implemented system by providing suitable data/results for comparison.

LDPC decoders using variants of Min-Sum algorithms is designed in [3]. The Min-Sum Unconditional Correction (UC) has changes only in the check node structure with a fixed correction factor subtracted from the magnitude of the outgoing messages of check nodes. In Min-Sum Successive Relaxation (SR) algorithm, the only change is in variable node architecture with a correction factor. A ½ rate 504-bit regular (3, 6) LDPC decoder using MS-UC is synthesized for Xilinx Virtex2 FPGA and shown to achieve a maximum throughput of 1.075 Gbps at 64 MHz. The decoder uses 4 bits word length with maximum of 15 iterations achieving BER of 10^{-5} at E_b/N_o of 3.4 dB. The throughput of the decoder is quite low for a parallel architecture and also due to selection of short code length.

An FPGA architecture using Stochastic decoding technique for LDPC code is discussed in [13]. In this design, a stochastic code is generated by using a comparator between the input channels data and a random number. This code is then fed to the variable nodes. A ½ rate 1024-bit regular fully-parallel LDPC decoder is synthesized on Virtex4 FPGA. The design uses 8-bit precision with 32 iterations that achieves a throughput of 706 Mbps at 212 MHz. The BER performance obtained from simulations is 10^{-6} at E_b/N_o of 3 dB. The author also claims that this design offers about 53% improvement in area of the FPGA compared to the fully-parallel approximate Min-Sum LDPC decoder architecture. Unfortunately, reliability is a concern for using stochastic decoding techniques.

A fully-parallel Stochastic LDPC decoder is presented in [2]. This work is an extension of [10] by the same authors. A ½ rate 1056-bit irregular LDPC decoder FPGA architecture is proposed in this paper. The design is synthesized on Xilinx Virtex4 FPGA. The maximum throughput of the implemented design is 1.66 Gbps at 222 MHz with 32 iterations. The BER performance is about 10^{-8} at 4.25 dB SNR. But, reliability is a concern for using stochastic based decoding techniques and also with an increased complexity in implementing stochastic sequence generator.

A bit-serial approximate Min-Sum LDPC decoder is proposed in [14]. The message communication between check nodes and variable nodes is done serially, in order to reduce interconnect complexity. Nodes in the decoder operate in an interlaced fashion to process frames simultaneously, without wasting the clock cycles. This mechanism doubles the throughput of the system. A 480-bit LDPC decoder is implemented on Altera Stratix2 FPGA. The maximum throughput achieved for 15 iterations per frame with 3-bit word length is about 650 Mbps at a maximum clock frequency of 61 MHz. However, the hardware requirement for implementation of serial-to-parallel/parallel-to-serial conversion of data still does not contribute enough to compensate the interconnect requirements of parallel architecture.

In [15], an FPGA based LDPC decoder using Layered Decoding algorithm has been designed. This design performs check node operation immediately after the variable node operations, and hence increasing the speed of the decoding operation. The design uses a Progressive Edge Growth (PEG) algorithm to generate a QC-LDPC code. A ½ rate 2304-bit code with 8 iterations is simulated, which achieves a BER of 10^{-6} at E_b/N_o of 2.1 dB. The design is synthesized for Stratix2 FPGA with 48% (69,112) of ALUTs, 17% (26,392) of flip-flops and 11% (1,084,416 bits) of RAM utilization. The authors claim to achieve a maximum throughput of 768 Mbps at 128 MHz clock frequency using this architecture. But this architecture requires custom LDPC matrix structure specifically for layered decoding and hence resulting in several limitations in code construction, decoding performance and implementation complexity.

In [16], an FPGA based flexible LDPC decoder to decode structured and unstructured code is designed. Min-Sum Algorithm is used in this design. The design uses memory banks to store the parity check matrix (H) information, hence making it flexible for different codes. The design uses partially-parallel architecture with fewer variable and check node processing units controlled by a controller unit. The regeneration of these blocks defines the parallelism in this architecture. The controller uses a mapping and scheduling algorithm to read H-matrix information and decode the data. A ½ rate 64,800-bit code with maximum of 15 iterations is implemented on Virtex4 FPGA to achieve a throughput of up to 11 Mbps at 74 MHz. But the throughput of the system is too low for high speed applications like DVB-S2.

In [17], a high throughput LDPC decoding architecture is designed using Min-Sum algorithm. The decoding architecture operates by scheduling variable and check node operations to process in a pipelined fashion keeping both the nodes busy at any point of time. The (7493, 6096) code is implemented on Stratix2 FPGA with a total resource utilization of 28,651 ALUTs, 22,012 registers and 1,406,828 memory bits, achieving a maximum throughput of 236.66 Mbps at 163.4 MHz clock frequency. However, the proposed solution utilizes massive memory resources.

A partially-parallel 9216-bit, ½ rate (3, 6) regular LDPC code has been designed in [18]. The proposed architecture uses memories to route the code information between variable and check node units. The design is synthesized for Virtex2 FPGA with a resource utilization of 46% (11,792) slices and 19% (10,105) registers. The decoder with a maximum of 18 iterations achieves a throughput of 54 Mbps at 56 MHz clock frequency from postrouting static timing analysis. The BER of 10^{-6} at E_b/N_o of 2 dB is calculated from simulation results. Although the BER performance is quite good, the operating frequency and throughput is not sufficient for high speed practical applications.

A high throughput Hierarchical Quasi-Cyclic (H-QC) LDPC decoder is proposed in [19]. A two-level regular H-QC LDPC code matrix structure is used in the decoder to parallelize row and column operations. A Scheduling algorithm is used in this design. The algorithm takes care of scheduling these operations to avoid memory access conflicts. The authors claim to have reduced memory requirement in this design compared to the pipelined decoders without scheduling. A multi-threaded architecture is implemented for ½ rate regular 12288-bit LDPC code on Altera Stratix2 FPGA. The throughput of 32-threaded partially-parallel LDPC decoder with 15 iterations is 298.1 Mbps at 96.26 MHz clock frequency. The BER is 10^{-5} at E_b/N_o of 1.55 dB from FPGA implementation. However, a multi-threaded architecture utilizes a huge hardware resource which is costly for practical applications.

Multi-Rate capability LDPC decoder architecture for FPGA is proposed by Lei Yang, et al., in [20]. The architecture is flexible for different rates and can be used for both regular and irregular LDPC codes as well. A 10k bits LDPC decoder is synthesized for Xilinx Vertex2 FPGA. For ½ rate irregular LDPC decoder, a maximum throughput of 30 Mbps at 100 MHz clock frequency is achievable from this design. At 60 iterations, the BER performance is 10^{-5} at 1.4 dB SNR. However, the throughput of the decoder is quite low for practical applications.

A modified Min-Sum algorithm with semi-parallel architecture for LDPC decoder is proposed in [21]. A scaling factor of 0.75 is used for the soft channel inputs to enhance the BER performance of the decoder. The check nodes and variable nodes are used 16 times in each of the iterations. Multiple dual-port memories are used to store intermediate messages in this architecture. A ½ rate (3, 6) 1536-bit LDPC decoder having 5-bit message precision and with a maximum of 20 iterations achieves a data rate of up to 127 Mbps at 121 MHz clock frequency. The design is implemented and tested on Xilinx Virtex2 FPGA and claims to achieve BER performance of 10^{-4} at 2.4 dB SNR. But, the architecture could have been better designed to operate at lower iterations and with a longer block length to offer even better throughput.

A quasi-cyclic LDPC decoder using Block Serial Scheduling technique is presented in [22]. The author claims that the proposed architecture reduces the need of message passing memory by 80 percent compared to standard message passing and reduction in routing requirements of about 50 percent as well. The 0.9 rate, 5490-bit LDPC decoder is implemented on Xilinx Virtex5 FPGA, achieving a throughput of about 1.27 Gbps at 153 MHz clock frequency. However, the analysis of BER performance of the decoder was not done or is unavailable for comparison.

A finite precision based LDPC decoder is discussed in [23]. Min-Sum algorithm is used in this design. The paper deals with Density Evolution to evaluate the performance variations in fixed point decoder implementations, with a variety of code rate and modulation scheme combinations. A ½ rate LDPC decoder is designed for QPSK, 16QAM and 64QAM channels. For QPSK modulation scheme, regular (3, 6) 9036-bit LDPC decoder implemented on Xilinx Virtex2 FPGA, a throughput of 60 Mbps with a maximum of 24 iterations is said to be achieved. But, the throughput of the decoder is quite low for high speed applications.

In [24], a modified Min-Sum decoding algorithm for LDPC codes based on Classified Correction is proposed. The proposed algorithm utilizes two corrections for both minimum and sub-minimum magnitudes of input messages in check nodes. These two correction factors can be obtained by analyzing the offset of updated messages in check nodes between the belief-propagation and the min-sum algorithms associated with check node degree. An 8176-bit LDPC code with word length of 7 bits is synthesized on Xilinx Virtex2 FPGA. With a maximum of 20 iterations, the decoder achieves a throughput of about 200 Mbps at 155.8 MHz clock frequency. However, this architecture requires design of complex check nodes.

A partially-parallel LDPC decoder using modified Min-Sum algorithm is presented in [25]. Euclidian Geometry (EG) based QC-LDPC codes are used in this design. The degree of parallelism incorporated in this design is 2. This work also compares performance of decoders with uniform and non-uniform quantization of message bits. The authors claim that the decoder can achieve a throughput of 170 Mbps at 193.4 MHz for 8176-bit code with a maximum of 15 iterations for decoding. Since this is partially-parallel architecture, the decoder requires 517 clock cycles per iteration. However, BER analysis results are unavailable for the proposed architecture to assess the overall performance of the decoder.

An irregular structured ½ rate 10240-bit LDPC decoder is proposed in [26]. A modified Log-Belief Propagation algorithm is used in this design. The proposed partially-parallel architecture has a parallelism degree of 2. The design is simulated with 5-bit fixed point message data with BPSK modulation in AWGN channel. The authors claim to achieve a maximum throughput of 223 Mbps based on the post PAR synthesis results of the design targeted for Xilinx Virtex II FPGA. But, the analysis of BER results which contributes to the overall performance of the decoder is not available for assessment.

In [27], a partially-parallel LDPC decoder is proposed using Min-Sum algorithm. The decoder is designed to comply with 802.16e WiMAX mobile standards. The design uses parallelism factor of 1, 4, and 6 degrees. The decoder is simulated for various code rates—1/2, 2/3, 3/4, and 5/6 and for code lengths of 576 and 2304 bits. The authors claim to have achieved a maximum throughput of 30 Mbps at 160 MHz clock frequency with 6 parallel blocks and a maximum of 20 iterations. However, the throughput of the proposed architecture is quite low for practical applications.

An LDPC decoder based on Joint Row-Column algorithm is designed in [28]. This algorithm processes each row of the parity check matrix one by one from the top row to the bottom row sequentially. This technique reduces the memory storage conflicts of extrinsic information generated by check nodes. In this work, 40 such joint processors are implemented to achieve a maximum throughput of 2 Gbps at a clock frequency of 150 MHz. Using parallel instantiations of decoding unit increases the throughput, but results in utilization of massive hardware resources including memory.

A memory efficient block serial architecture is proposed in [29] for multi-rate and multi-length irregular LDPC decoder. The Approximate-Min (A-Min) algorithm is used in this paper. This architecture is designed to comply with IEEE 802.16 Wireless MAN standard. The decoder targeted on Altera FPGA is claimed to operate at 160 MHz with a maximum of 15 iterations and 8-bit word length. However, the throughput of the decoder is not available to ascertain the performance of the decoder.

A low memory FPGA based LDPC decoder architecture for Quasi-Cyclic LDPC codes is proposed in [30]. A modified Turbo decoding algorithm is used in this work. Quasi-Cyclic LDPC codes have been used to reduce interconnect complexity. The design is simulated in software as well as verified on Xilinx Virtex 4 FPGA. A comparative analysis of LDPC decoder architecture performance is done with respect to BER and memory utilization against other implementations. The throughput of the decoder is not available to ascertain the performance of the decoder.

A partially-parallel LDPC decoder using Belief Propagation algorithm is presented in [31]. It is also shown by simulations that the BER performance improves as the code length increases. The quantization effect against floating point precision of soft input data is also analyzed. A non-uniform quantization technique is used in this design. In this partially-parallel architecture, a (32, 16) decoder block is replicated 128 times to form (4096, 2048) decoder that achieves a maximum throughput of 31 Mbps with 10 fixed iterations on Virtex II FPGA. However, the throughput is quite low for practical applications.

A partially-parallel LDPC Decoder using Message-Passing algorithm is proposed in [8]. The decoder performs the column operations for variable nodes in conjunction with the row operations for check nodes. A ½ rate regular (3, 6) 1536-bit LDPC decoder with 8-bit quantization is synthesized for Xilinx Virtex II FPGA. The maximum throughput of 38 Mbps is achievable at a maximum clock frequency of 100 MHz with BER performance of 10^{-5} at 4 dB SNR. However, the throughput of the decoder is quite low for practical applications.

In [32], FPGA architecture for LDPC coding that allows code specific architecture; also providing dynamic code selection is described. The code is stored in the ROM and the microcontroller loads the FPGA with the selected code architecture. A ½ rate 1008-bit LDPC decoder is synthesized for Virtex4 FPGA. The LUT counts for different message widths against different degrees of check nodes which are also compared in this paper. However, this implementation is based on microcontroller-FPGA architecture and cannot make a fair comparison with a fully FPGA based architecture in terms of hardware resources.

An FPGA based unstructured LDPC decoder is proposed in [5]. The decoder uses a serial architecture. The decoder is connected to the PowerPC Core by the Processor Local Bus (PLB) on the FPGA. The address map of the decoder is controlled by the PowerPC. This allows the flexibility of programming different prototypes of the decoder. This design uses a modified Sum-Product algorithm, a class of Min-Sum algorithm with correction factor. The proposed design is synthesized on Xilinx Vertex2 FPGA for 0.82 rate 4095-bit LDPC code with 8-bit word length. With a maximum iteration of 10 cycles, the decoder achieves a throughput of 2 Mbps and BER of 10^{-6} at 4 dB SNR. But, using processor-based implementation hampers the throughput of the decoder.

A comparison of the design complexity and performance of various LDPC decoder architectures is shown in Table 5.1. It is noted that with an increased design complexity, partially-parallel architecture provides a trade-off between the

Table 5.1 Design Complexity and Performance of LDPC Decoder Architectures

Decoder Architecture	Fully Parallel	Fully-Serial	Partially-Parallel
Design effort	Simple	Moderate	Complex
Hardware requirement	High	Low	Medium
Routing message edges	Complex	Simple	Moderate
Block memory utilization	No	Yes	Yes
Memory access conflict	Not Applicable	Exists	Exists
Average throughput	High	Low	Moderate

hardware requirement and throughput compared to fully parallel and fully-serial architectures. The other factors listed in the table are also reasonable in the case of partially-parallel architecture.

5.3 SUMMARY

An exhaustive literature review has been carried out based on the publications right from the time when LDPC codes gained popularity (year 2000 onwards). The literature review is mainly focused on the implementation of LDPC decoders on FPGA, particularly on various decoding algorithms, hardware architectures, design specifications, application standards, and performance parameters. The key parameters of the reviewed decoders are tabulated for comparison of their performance. The comparison table is available in Appendix A. From the table, it is observed that some of the partially-parallel decoders using Min-Sum based algorithm can achieve throughput in the range of 10 to 250 Mbps [17,27]. The throughput can be increased by using Layered Decoding algorithm, achieving a range of 38 Mbps to 2 Gbps [8,28]. Fully Parallel decoders offer much higher throughput (600 Mbps to 6 Gbps) using Min-Sum or Stochastic based decoders at the expense of higher hardware resources [10,13].

Based on the literature review on different LDPC decoding algorithms and hardware architectures, it can be observed that the majority of the decoder implementations are based on the Min-Sum algorithm a modified form of the Sum-Product algorithm. This is because of the reduced complexity of node operations in Min-Sum algorithm, making it very suitable for hardware implementations. The BER performance of the Min-Sum algorithm is close to that of the Sum-Product algorithm. Hence Min-Sum algorithm has been the most popular choice for designing LDPC decoders to date. An important aspect that can be noted from the literature review is that the majority of the decoder designs have only been synthesized and not practically implemented or tested on hardware. It indicates that the practical implementation of LDPC decoders is still a challenging task due to the complexity involved in the design, verification, and testing process.

In [10], a fully-parallel architecture using the Min-Sum LDPC decoding algorithm is synthesized for $\frac{1}{2}$ rate (3,6) regular 1200-bit code with 3-bit LLR quantization. Using a maximum of 10 decoding iterations, the decoder achieves a throughput of 6 Gbps with a BER of 10^{-4} at E_b/N_o of 3.75 dB. This design requires massive hardware resources, utilizing the majority of the slices on Xilinx Virtex4 FPGA while rest of the chip is occupied by the routing logic. This example clearly indicates that 1200-bit LDPC code can hardly be synthesized using fully-parallel architecture, making it impossible for implementation of larger codes on a Virtex4 FPGA.

A serial LDPC decoder using Min-Sum algorithm for 0.8 rate irregular 4095-bit code with 8-bit LLR quantization has been synthesized in [5]. With a maximum of 10 decoding iterations, the throughput achievable is 2 Mbps at a BER of 10^{-6} at

E_b/N_o of 4 dB. This design utilizes about 6% of the slices on a Xilinx Virtex2 FPGA. Although the hardware resources utilized by this implementation is significantly reduced, the throughput achieved is very low. Therefore, the decoder performance is not on par with the requirements of practical high speed applications.

In [8], a partially-parallel architecture of LDPC decoder using Min-Sum algorithm for ½ rate (3,6) regular 1536-bit code with 8-bit LLR quantization has been synthesized for an FPGA. Using a maximum of 7 decoding iterations the decoder achieves a throughput of 38 Mbps and a BER of 10^{-5} at E_b/N_o of 4 dB. This design utilizes about 4% of the slices and 70% of the block RAM on a Xilinx Virtex2 FPGA. Clearly, the throughput improvement in partially-parallel architectures over serial architectures is only a few Mbps and well below the performance of fully-parallel architectures. The hardware resources required by this architecture are significantly less compared to fully-parallel architectures. However, careful designing of the memory control block is needed to avoid memory access conflicts.

From the above discussion, it can be concluded that the challenges of hardware implementation of LDPC decoders are not just with the architectural design, but there is a need for low complexity decoding algorithms as well. It is obvious that simplification of the decoding algorithm helps with efficient hardware implementation and to achieve higher performance in terms of chip area and throughput. However, the simplification of algorithm may result in the degradation of BER performance. Achieving a balanced trade-off in terms of (1) BER performance, (2) logic/memory requirements and throughput, and (3) complexity of implementation and verification of the decoder on hardware is a challenging problem. The remaining chapters of this book will explore these issues in depth and suggest possible alternatives.

REFERENCES

[1] V.A. Chandrasetty, S.M. Aziz, An area efficient LDPC decoder using a reduced complexity min-sum algorithm, Integr. VLSI J. 45 (2) (2012) 141−148.

[2] S. Sharifi Tehrani, S. Mannor, W.J. Gross, Fully parallel stochastic LDPC decoders, IEEE Trans. Signal Process. 56 (11) (2008) 5692−5703.

[3] S. Tolouei, A.H. Banihashemi. FPGA implementation of variants of min-sum algorithm. in 24th Biennial Symposium on Communications, Kingston, ON, 2008.

[4] S. Bates, et al., A low-cost serial decoder architecture for low-density parity-check convolutional codes, IEEE Trans. Circuits Syst. I Regul. Pap. 55 (7) (2008) 1967−1976.

[5] S.M.E. Hosseini, , C. Kheong Sann, G. Wang Ling. A reconfigurable FPGA implementation of an LDPC decoder for unstructured codes. in: 2nd International Conference on Signals, Circuits and Systems, Monastir, 2008.

[6] S.M. Kim, C.S. Park, S.Y. Hwang, A novel partially-parallel architecture for high-throughput LDPC Decoder for DVB-S2, IEEE Trans. Consumer Electr. 56 (2) (2010) 820−825.

[7] V.A. Chandrasetty, S.M. Aziz. A multi-level hierarchical quasi-cyclic matrix for implementation of flexible partially-parallel LDPC decoders. in IEEE International Conference on Multimedia and Expo, Barcelona, Spain, 2011.

[8] K. Shimizu, et al. Partially-parallel LDPC decoder based on high-efficiency message-passing algorithm. in: IEEE International Conference on Computer Design: VLSI in Computers and Processors, 2005.

[9] D. Yongmei, C. Ning, Y. Zhiyuan, Memory efficient decoder architectures for quasi-cyclic LDPC codes, IEEE Trans. Circuits Syst. I Regul. Pap. 55 (9) (2008) 2898–2911.

[10] R. Zarubica, S.G. Wilson, E. Hall. Multi-Gbps FPGA-based low density parity check (LDPC) decoder design. in: IEEE Global Telecommunications Conference. Washington, DC, 2007.

[11] W.J. Gross, V.C. Gaudet, A. Milner. Stochastic Implementation of LDPC Decoders. in: 39th Asilomar Conference on Signals, Systems and Computers, Pacific Grove, CA, 2005.

[12] D.A. Morero, G. Corral-Briones, M.R. Hueda, Parallel architecture for decoding LDPC Codes on high speed communication systems, Argentine School of Micro-Nanoelectronics, Technology and Applications, Buenos Aires, 2008.

[13] S.S. Tehrani, S. Mannor, W.J. Gross. An area-efficient FPGA-based architecture for fully-parallel stochastic LDPC decoding. in: IEEE Workshop on Signal Processing Systems, Shanghai, China, 2007.

[14] A. Darabiha, A.C. Carusone, F.R. Kschischang. A bit-serial approximate min-sum LDPC decoder and FPGA implementation. in: IEEE International Symposium on Circuits and Systems, Island of Kos, Greece, 2006.

[15] H. Ding, et al. Design and implementation for high speed LDPC decoder with layered decoding. in: WRI International Conference on Communications and Mobile Computing, Yunnan, 2009.

[16] C. Beuschel, H.J. Pfleiderer. FPGA implementation of a flexible decoder for long LDPC codes. in: International Conference on Field Programmable Logic and Applications, Heidelberg, 2008.

[17] Y. Zhixing, et al. High-throughput LDPC decoding architecture. in: International Conference on Communications, Circuits and Systems, Fujian, 2008.

[18] Z. Tong, K.K. Parhi. A 54 Mbps (3,6)-regular FPGA LDPC decoder. in: IEEE Workshop on Signal Processing Systems, 2002.

[19] C. Yi-Hsing, K. Mong-Kai. A high throughput H-QC LDPC decoder. in: IEEE International Symposium on Circuits and Systems, New Orleans, LA, 2007.

[20] L. Yang, et al., An FPGA implementation of low-density parity-check code decoder with multi-rate capability, Conference on Asia South Pacific Design Automation, ACM, Shanghai, China, 2005.

[21] M. Karkooti, J.R. Cavallaro. Semi-parallel reconfigurable architectures for real-time LDPC decoding. in: International Conference on Information Technology: Coding and Computing, 2004.

[22] K.K. Gunnam, et al. Decoding of quasi-cyclic LDPC codes using an on-the-fly computation. in: 40th Asilomar Conference on Signals, Systems and Computers, Pacific Grove, CA, 2006.

[23] S. Manyuan, et al. Finite precision implementation of LDPC coded M-ary modulation over wireless channels. in: 37th Asilomar Conference on Signals, Systems and Computers, 2003.

[24] Z. Zhou, et al. Modified min-sum decoding algorithm for LDPC codes based on classified correction. in: 3rd International Conference on Communications and
Networking, Hangzhou, 2008.

[25] C. Zhiqiang, W. Zhongfeng. A 170 Mbps (8176, 7156) quasi-cyclic LDPC decoder
implementation with FPGA. in: IEEE International Symposium on Circuits and
Systems, Island of Kos, Greece, 2006.

[26] W. Wang, et al. A 223Mbps FPGA implementation of (10240, 5120) irregular structured low density parity check decoder. in: IEEE Vehicular Technology Conference,
Singapore, 2008.

[27] F. Charot, et al. A new powerful scalable generic multi-standard LDPC decoder
architecture. in: 16th International Symposium on Field-Programmable Custom
Computing Machines, Palo Alto, CA, 2008.

[28] H. Zhiyong, R. Sebastien, F. Paul. FPGA implementation of LDPC decoders based
on joint row-column decoding algorithm. in IEEE International Symposium on
Circuits and Systems, New Orleans, LA, 2007.

[29] Z. Xiyu, Z. Zhaoyang. Memory efficient block-serial architecture for programmable,
multi-rate multi-length LDPC decoder. in: 2nd International Conference on
Communications and Networking, Shanghai, China, 2007.

[30] P. Saunders, , A.D. Fagan. A low memory FPGA based LDPC decoder architecture
for quasi-cyclic LDPC codes. in: IET Irish Signals and Systems Conference, Dublin,
2006.

[31] J. Lee, et al. A scalable architecture of a structured LDPC decoder. in: IEEE
International Symposium on Information Theory, Chicago, USA, 2004.

[32] O.J. Hernandez, N.F. Blythe. An FPGA architecture for low density parity check
codes. in IEEE Southeastcon, Huntsville, AL, 2008.

Hardware implementation of LDPC decoders

6

6.1 DECODER DESIGN METHODOLOGY

This chapter presents the architectural design and field programmable gate array (FPGA) implementation of LDPC decoders based on the reduced complexity algorithms discussed in Chapter 4, LDPC decoding algorithms (SMP and MMS). Fully-parallel and partially-parallel decoder architectures for both of the algorithms are discussed. It is known that fully-parallel implementation of LDPC decoders requires a large amount of hardware resources compared to partially-parallel architectures. However, the primary motivation for fully-parallel implementation is to ascertain the performance as well as the savings in hardware resources when using the reduced complexity algorithms presented in this book. This information can be used to develop suitably partitioned partially-parallel architectures to achieve further savings in hardware resources while providing acceptable level of BER performance [1]. In addition, there are many applications which can still accommodate a fully-parallel decoder of small- to medium-code length and benefit from the high throughput [2,3]. Finally, the low-complexity algorithms presented in this book can be incorporated into some of the existing partially-parallel architectures [4−9] to reduce their hardware resource requirements further. An area and memory efficient partially-parallel LDPC decoder architecture has been designed and implemented on FPGA. This architecture is very flexible in supporting multiple code lengths for WLAN and LTE applications. The subsequent sections present the details of the architectural design and performance analysis of fully-parallel and partially-parallel decoders practically implemented as part of this research.

6.1.1 DESIGN AND IMPLEMENTATION

The high level block diagram of an LDPC decoder depicting the hardware interfaces that are common to both fully-parallel and partially-parallel architectures is shown in Fig. 6.1. The decoder consists of a global *Clock* and synchronous "Reset" inputs. The maximum permissible number of iterations is determined by the value supplied at the *MaxIter* input. When the *Configure* input is high, the *MaxIter* value is read. The LLRs are fed into the decoder using the *Load* control signal. The decoding process is initiated by the *Start* signal. After the decoding is complete, the *Decoded*

Resource Efficient LDPC Decoders. DOI: https://doi.org/10.1016/B978-0-12-811255-7.00006-X

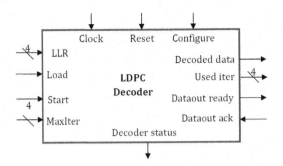

FIGURE 6.1

Hardware interfaces of the LDPC decoder.

Data can be obtained when indicated by the *DataOut Ready* signal. The receipt of data can be acknowledged on *DataOut Ack* to receive the next decoded bit. The number of iterations used for decoding can be obtained from *Used Iter* port. The *Decoder Status* port indicates the progress (Active/Idle) of the decoder.

Note that the LLRs are loaded serially (one at a time) into the decoder. Similarly, the *Decoded Data* is latched bit by bit serially. This technique is used because of the limited number of Input/output ports available in the FPGA. It also provides flexibility for implementing LDPC decoders with variable code lengths without the need to modify the port configuration.

A parameterized Register Transfer Level (RTL) model of the decoder has been developed using Verilog Hardware Description Language (HDL) [10] and synthesized using Xilinx Synthesis Tool. Test bench models were developed to verify the functionality of different modules and sub-modules of the decoder. Simulation of behavioral, posttranslate, and post Placement and Route (PAR) models were carried out using ModelSim. The procedure used for modeling, design, and simulation of the LDPC decoder is illustrated in Fig. 6.2.

An LDPC code for practical applications varies in length and sparseness of non-zero elements in the matrix. The code length for WLAN standard typically varies from 648 to 1944 bits, and for LTE, it is in the order of 576 to 2304 bits. Although the variable and check nodes in the decoder are identical and repetitive, the inter-connections between the nodes make the structure highly complex. Hardware design of such complex codes is a cumbersome task for the designer [11]. In the case of fully-parallel architecture, a huge amount of time is required to model the thousands of interconnections in HDL. Simulating and debugging the hand coded HDL model is an extremely difficult and time-consuming process. Modeling the decoder with a different code length, or for another application, becomes an unproductive and repetitive activity. Therefore, an efficient design methodology is necessary to automate the design flow to eliminate redundant and repetitive tasks.

An automation tool has been developed using the C programming language and executed in the MATLAB environment to address the above issues. This

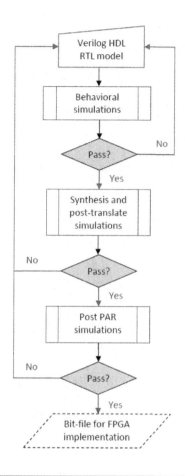

FIGURE 6.2

Flow chart illustrating the design and simulation procedure for LDPC decoder.

application is able to generate variable inter-connect bit-widths (extrinsic message length) as specified by the user, particularly for simplified algorithms (SMP and MMS). The behaviors of the variable node and check node modules are hand coded according to the algorithms, because these modules are identical throughout a design and do not require huge effort. An illustration of the automation technique developed for fully-parallel architecture is shown in Fig. 6.3A.

A similar technique has been used for partially-parallel architectures. Particularly, the 3L-HQC-LP matrix presented in Chapter 3, Structure and flexibility of LDPC codes, requires storing information on permuted random matrices in the LUTs of the decoder. This information can be systematically obtained from the LDPC matrix and hence can be automated to generate appropriate HDL code in LUT format. Depending on the parallelism factor of the decoder, an HDL

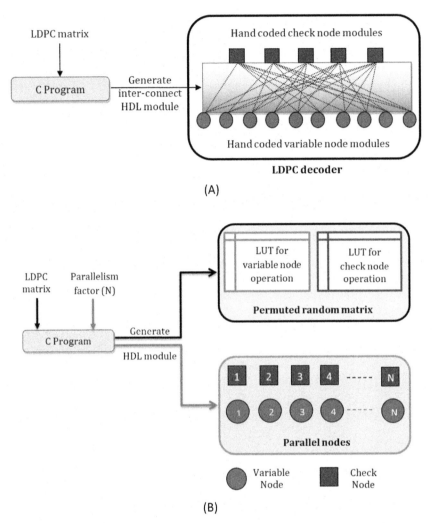

FIGURE 6.3

Illustration of the design automation technique for LDPC decoders. (A) Fully-parallel decoder architecture, (B) Partially-parallel decoder architecture.

module for parallel nodes can also be generated. An illustration of the automation technique developed for partially-parallel architectures is shown in Fig. 6.3B.

The above automation technique can generate appropriate HDL code for a given LDPC matrix, thereby simplifying the task of prototyping various algorithms, architectures, and LDPC code lengths. The MATLAB code for generating the matrix is available in Appendix A. The technique can save significant design effort and time.

6.1.2 PERFORMANCE MEASUREMENT

1. *Interconnect routing complexity (I)*: The routing complexity of the algorithms is assessed by calculating the number of node inter-connects (I) in the decoder using the formula shown in Eq. 6.1. This parameter is used to evaluate the hardware implementation complexity particularly of the fully-parallel decoder architectures.

$$I = 2 \times C_l \times E_l \times V_d \qquad (6.1)$$

   where C_l is the code length, E_l is the extrinsic message length and V_d is the degree of the variable node. The calculation includes the interconnections from variable node to check node and vice versa, by multiplying the equation by factor *2* in the formula.

2. *Clocks per decoding iteration (CDI)*: In certain hardware implementations of LDPC decoder, each of the node processing is carried out at several stages with in one decoding iterations. Hence it necessitates the requirement of additional clock cycles for processing these stages as well as to compensate the latency in such multistage systems. This parameter significantly impacts the throughput of the decoder.

3. *Decoder throughput (T):* The throughput of the decoder is calculated using the formula shown in Eq. 6.2 [12]. This calculation excludes the serial load time of individual LLRs (before starting the decoding process) and latch time of decoded data (after decoding is complete). This parameter is commonly used to evaluate the speed of operation of both fully-parallel and partially-parallel decoder architectures.

$$T = \frac{r \times C_l \times f_{max}}{N_{it} \times \eta} \qquad (6.2)$$

   where, r is the rate and C_l is the length of the LDPC code, f_{max} is the maximum operating frequency, N_{it} is the number of decoding iterations and η is the number of clock cycles required to complete one iteration (CDI). $\eta = 1$ for fully-parallel architecture implementation of the decoder.

4. *RAM memory bits (M_B)*: This parameter measures the total memory requirement of the decoder. The memory in bits utilized by the complete decoder in the form of Block RAMs (BRAM) is used to assess the overall performance of the decoder in terms of area. This parameter is particularly associated with partially-parallel decoder architectures, or in some cases where BRAMs are utilized.

6.2 PROTOTYPING LDPC CODES IN HARDWARE

Any hardware design that is intended for practical applications requires implementation of prototype models on the hardware for testing [13]. Field Programmable

Gate Array (FPGA) is a very flexible and widely used platform for rapid prototyping, and is also a low-cost approach compared to ASIC implementation.

FPGA is an integrated circuit that contains large numbers of identical logic cells that can be interconnected by a matrix of wires using programmable switch boxes [14]. A design can be implemented by specifying the simple logic function for each cell and selectively closing the switches in the interconnect matrix. The array of logic cells and the mesh of interconnecting wires form the basic building block of an FPGA. Complex designs can be implemented by programming these basic building blocks [15].

FPGAs offer a number of benefits over other implementation flows such as ASIC and off-the-shelf DSP and microcontroller chips. Some of the benefits of using FPGA are as follows [16,17]:

1. *Performance*: FPGAs offer logic structures that provide the advantage of incorporating parallelism in designs and thereby significantly enhance the computational speed compared to processor-based platforms.
2. *Reliability*: Processor-based designs operate on instructions to perform a particular task using shared hardware resources. However, FPGA-based designs consist of dedicated hardware for performing such tasks with predictable delays. Hence, increasing the reliability of real-time systems.
3. *Long-term maintenance*: FPGAs provide flexibility in upgrading the design in case of a change in specification of an application over time. The time spent in redesigning/enhancing a FPGA-based design is much less compared to that of ASIC design.
4. *Cost*: The Non-Recurring Engineering (NRE) cost for designing a custom ASIC is huge compared to FPGA-based solutions. FPGAs bypass the backend physical design, fabrication, and packaging cost as well.
5. *Time to Market*: FPGA technology provides flexibility for rapid prototyping of the design by avoiding fabrication and other processing delays, thus facilitating quicker "time to market" solutions.

The major limitation of FPGA-based design is in its overall performance compared to ASIC solutions. However, it provides the best methodology for quick prototyping and testing of the designs including the above listed advantages.

Today's FPGAs are capable of implementing a wide range of diverse functions, for example Digital Signal Processors (DSP), networking and routing processors, communication processors, image and video processors, number crunchers, and many more. DSP building blocks such as filters, transforms, multipliers, comparators, etc., can be easily incorporated into a FPGA [18,19]. Also, the parallel processing and pipe-lining features inherent in FPGA are very useful in designing complex communication modules [20,21]. They are widely used for implementation of modems, error correction codecs, and multimedia processing applications [22−25]. To accelerate the design process, some of these modules are readily available as IP cores from popular vendors, namely, Xilinx [26], Altera [27], Actel [28], among others. Using FPGAs for designing hardware has proved to be one of the best approaches to balance and reduce the time required

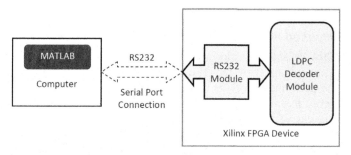

FIGURE 6.4

Block diagram of FPGA test setup using RS232 serial communication.

for practical implementation and prototyping. Hence, a FPGA-based solution is chosen for implementing the LDPC decoders in this book. A comprehensive literature review on FPGA-based LDPC decoders is presented in the next section.

The LDPC decoders were implemented on a Xilinx Virtex5 FPGA (XC5VLX110T). A comprehensive testing environment was developed to test the implemented decoders. The FPGA test setup using RS232 serial communication is shown in Fig. 6.4. To interface with the RS232 port of the computer, an RS232 transceiver module was embedded on the FPGA along with the LDPC decoder module [29]. MATLAB was used to communicate with the FPGA. A serial port communication driver was developed using the C programming language and executed in the MATLAB environment [30]. A maximum baud rate of 115200 was used for the serial data communication. To begin with the testing process, the LLRs are first sent to the FPGA along with the appropriate control signals. After the decoding is complete in the FPGA, the decoded data is sent via the same serial port to the computer to analyze the performance of the decoder. The number of iterations used for decoding is also collected to estimate the average throughput of the implemented decoder. The same sets of LLRs are used for simulation of the software model of the decoder in MATLAB. This technique is used to compare and verify the BER performance of the hardware decoder on FPGA to that of the software model under similar conditions. A snapshot of the FPGA development board used in this research for implementing and testing the LDPC decoder is shown in Appendix B.

6.3 IMPLEMENTATION OF HARDWARE EFFICIENT DECODER

As stated previously, fully-parallel implementation of decoders for large LDPC codes has been problematic due to the large amount of resources required. This section presents fully-parallel LDPC decoder architectures based on the Simplified Message Passing (SMP) and Modified Min-Sum (MMS) algorithms (see Chapter 4: LDPC decoding algorithms) to assess their attractiveness for hardware implementation compared to other decoders.

6.3.1 FULLY-PARALLEL ARCHITECTURE

6.3.1.1 Simplified message passing decoder

A fully-parallel architecture using the SMP algorithm with ½ rate (3, 6) 648-bit LDPC code compliant with the WLAN standard (IEEE 802.11 n) has been designed. The performance of the decoder implemented on FPGA was analyzed and compared to that of the software model of the same decoder. The BER and FER performances of the implemented decoder are shown in Figs. 6.5 and 6.6 respectively. The loss in BER performance of the FPGA-based decoder is negligible compared to the software-based decoder and the loss in FER performance is less than 0.1 dB (E_b/N_o). The average number of iterations required by the FPGA-based decoder closely follows the average iterations predicted by the software simulation model, as shown in Fig. 6.7.

The summary of FPGA device utilization of the LDPC decoder is shown in Table 6.1. To the best of the authors' knowledge, FPGA implementation results for decoders based on the Bit-Flip (BF) and Sum-Product (SP) algorithms are not available in the literature for comparison with the SMP-based decoder. Hence, the

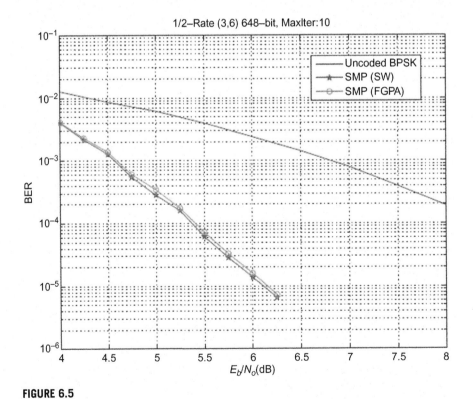

FIGURE 6.5

BER performance of the fully-parallel SMP decoder implemented on FPGA.

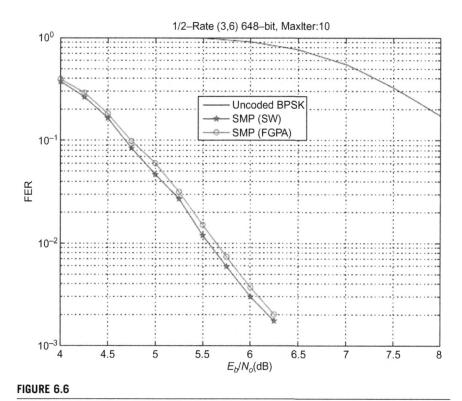

1/2–Rate (3,6) 648–bit, MaxIter:10

Legend:
— Uncoded BPSK
—★— SMP (SW)
—○— SMP (FGPA)

X-axis: E_b/N_o(dB)
Y-axis: FER

FIGURE 6.6

FER performance of the fully-parallel SMP decoder implemented on FPGA.

results for BF and SP were obtained from postplacement and routing (PAR) of the design. Note that for SP, only synthesis and mapping was carried out, since the Xilinx synthesis tool failed to route the design completely due to huge complexity.

Using the average decoding iterations from Fig. 6.7, the estimated average throughput of the LDPC decoder implemented on FPGA is computed using Eq. 6.2. The average throughput is found to be $\sim$16.2 Gbps at E_b/N_o of 6.25 dB. Fig. 6.5 shows that the BER achieved at this E_b/N_o is approximately 10^{-5}.

6.3.1.2 Modified Min-Sum decoder

The Modified Min-Sum (MMS) algorithm requires mapping of messages at the variable node, as discussed in Section 4.2.2. From a hardware implementation point of view, the mapping is to be performed for converting 4-bit messages to 2-bit values and vice versa. Fig. 6.8A shows the input and output messages of the variable node for the MMS decoder. A typical variable node structure is depicted in Fig. 6.8B, where the required message mapping is achieved using Look-Up Tables (LUT). This example shows a variable node of degree 3, i.e., it takes three

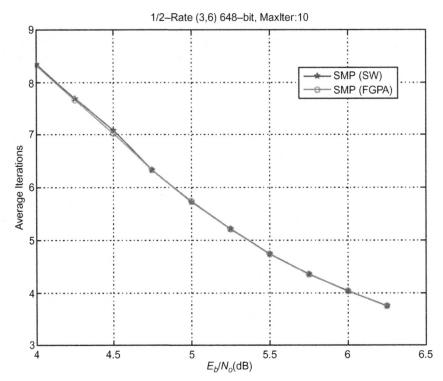

FIGURE 6.7

Average iterations of the fully-parallel SMP decoder implemented on FPGA.

Table 6.1 FPGA Device Utilization Summary

Resources	SP	SMP	BF
FPGA device	Xilinx Virtex 5(XC5VLX110T-3FF1136)		
LDPC code	½ rate (3, 6) regular 648-bit (WLAN)		
Intrinsic message (LLR)	4-bit	4-bit	1-bit
Extrinsic message	4-bit	1-bit	1-bit
Slices	15684	4046	1396
LUTs	58787	14239	3577
Registers	12443	5963	2069
Clock (MHz)	128	188	190

inputs and generates three outputs to the respective check nodes. Clearly, the additional LUTs required for message mapping lead to some hardware overhead in the MMS. It is possible to reduce this overhead by merging the LUTs required for message mapping and the addition/subtraction block that are normally required by the algorithm. Unless this merging is done at the design level, the synthesis tool uses separate LUT resources for message mapping and addition/subtraction. The resulting optimized variable node is shown in Fig. 6.8C.

The optimized and un-optimized variable nodes for MMS have been synthesized to estimate the difference in hardware resource requirements. The synthesis

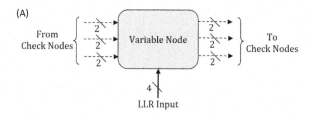

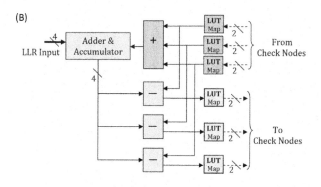

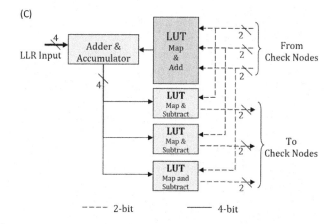

---- 2-bit ——— 4-bit

FIGURE 6.8

LUT-based variable node structures. (A) Input and output messages of the variable node for the MMS decoder, (B) Un-optimized variable node structure, (C) Optimized variable node structure.

Table 6.2 Synthesis Results of Variable Nodes for MMS Algorithm

Resources	Un-Optimized	Optimized	Improvement
Slices	22	8	64%
LUTs	49	26	47%
Registers	11	11	–
Clock (MHz)	456	392	−14%
FPGA	Xilinx Virtex5 (XC5VLX110T-3FF1136)		

Table 6.3 FPGA Implementation Results for MMS Based Fully-Parallel Decoder

Resources	Un-optimized	Optimized	Improvement
Slices	17,114	10,823	37%
LUTs	68,049	39,024	43%
Registers	15,107	15,107	–
Clock (MHz)	84.5	82	−7%
Avg. throughput (Gbps)	7.7	7.1	−7%
Total power (W)	1.45	1.13	22%
FPGA	Xilinx Virtex 5 (XC5VLX110T-3FF1136)		

results are shown in Table 6.2. It is clear that the variable node optimization presented above leads to significant reduction (64%) in FPGA slices required. However, the operating frequency is reduced by 14% in the optimized design due to longer chain of large LUTs in the logic.

A fully-parallel LDPC decoder has been implemented using the MMS algorithm (configuration C1, as defined in Section 4.3) with variable node optimization. The decoder implemented is a ½ rate (3, 6) regular 1152-bit LDPC code that is compliant with the LTE standard.

The results obtained from testing the LDPC decoder on FPGA are summarized in Table 6.3. The optimized decoder (where the variable node is optimized) requires 37 percent less FPGA Slices and saves up to 22% of total power compared to the un-optimized version. However, there is approximately 7% reduction in maximum operating frequency and throughput compared to the un-optimized version for reasons stated previously. This penalty is negligible when compared to the significant savings achieved in the hardware resources and power.

Various performance results obtained from tests conducted on the MMS decoder implemented on FPGA are shown in Figs. 6.9–6.11. Figs. 6.9 and 6.10 illustrate that both the BER and FER performances of the MMS decoder on

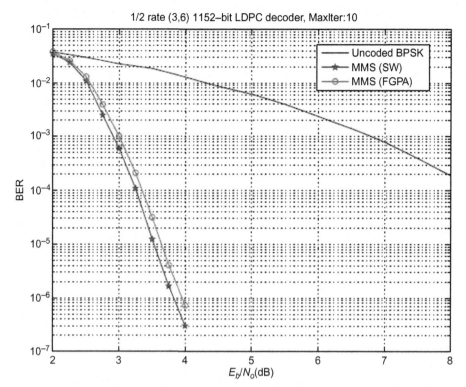

FIGURE 6.9

BER performance of the fully-parallel MMS decoder.

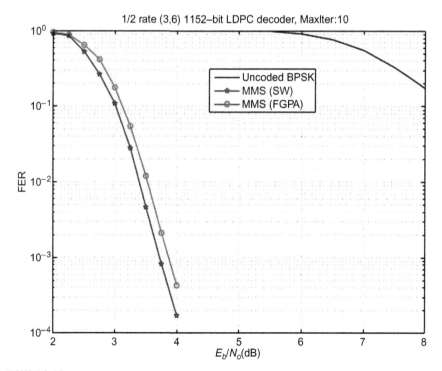

FIGURE 6.10

FER performance of the fully-parallel MMS decoder.

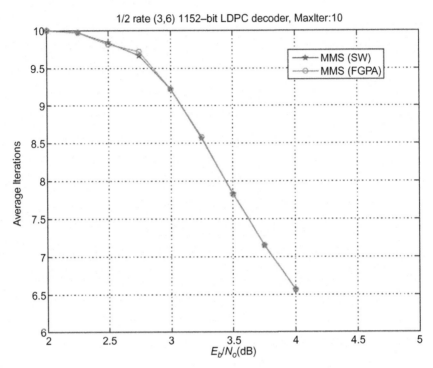

FIGURE 6.11

Average decoding iterations of the fully-parallel MMS decoder.

FPGA suffer a slight loss of about 0.15 dB compared to those of its software simulation model. This difference is due to the approximations made in the values stored in the reorganized LUTs for hardware optimization of the variable nodes. Fig. 6.11 shows that the average iteration for the FPGA-based MMS decoder closely follows that of the software-based decoder. Using the average iteration count of 6.6 at E_b/N_o of 4 dB (from Fig. 6.11) and a clock frequency of 82 MHz (from Table 6.3), the average throughput of the FPGA-based decoder can be estimated to be ~7.1 Gbps.

Some papers have reported synthesis of LDPC decoders using stochastic algorithms [31,32]. In stochastic-based decoders, the extrinsic messages are exchanged between the decoding nodes in a bit-serial fashion [33]. One iteration at the bit-level is called a *decoding cycle*. The termination of the algorithm is dependent on the maximum number of bit-level decoding cycles [34]. This is why the stochastic decoders presented in [31,32] require very large numbers of decoding cycles, as shown in Table 6.4, compared to a maximum iteration count of only 10 used by the MMS decoder. Table 6.4 also shows that the MMS decoder can achieve a much higher throughput compared to those in [31] and [32]. With comparable FPGA resource requirement, the MMS decoder suffers a

Table 6.4 Comparison of Fully-Parallel LDPC Decoders

Parameters	Modified Min-Sum (MMS)	2-bit Min-Sum	[12]	[31]	[32]
LDPC code	1152	1152	1200	1024	(1056, 528)
Rate & Regularity	½ rate (3, 6) Regular				½ rate Irregular
Algorithm	MMS (C1)	Min-Sum		Stochastic	
Intrinsic message	4-bit	2-bit	3-bit	8-bit	6-bit
Extrinsic message	2-bit	2-bit	3-bit	8-bit (1-bit serial)	6-bit (1-bit serial)
No. of inter-connects	13,824	13,824	21,600	6,144	–
Slices	33,345	31,142	40,613	32,875	46,097
LUTs	58,053	52,689	69,038	47,104	68,112
Registers	15,691	13,355	18,945	20,582	44,458
Clock (MHz)	123	123	100	212	222
Max. decoding iterations or cycles	10 decoding iterations			6000 decoding cycles	700 decoding cycles
E_b/N_o (dB) at BER of 10^{-6}	3.75	6.15	3.85	3.0	2.9
Avg. throughput at BER of 10^{-6}	~10.5 Gbps	~14.2 Gbps	~9.1 Gbps	~353 Mbps at BER of 10^{-6}	~470 Mbps at BER of 10^{-5}
FPGA device	Xilinx Virtex 4 (xc4vlx200)				

BER degradation of only about 0.6 dB at a BER of 10^{-6} compared to [31] and [32]. The BER performance of the stochastic decoders [31,32] are better due to the very large number of decoding cycles, the latter adversely affecting the throughput as stated above. Another issue with the stochastic decoders is the overhead associated with buffering long extrinsic serial messages. Table 6.4 shows that compared to 3-bit MS based decoder [12], the MMS decoder has a very small loss in BER performance (only 0.1 dB loss) while offering significant savings in hardware resources (about 18 percent less slices). Table 6.4 also reveals that compared to a 2-bit MS decoder the MMS decoder has a massive improvement in performance (2.4 dB improvement) at a BER of 10^{-6} with similar hardware requirement.

6.3.2 PARTIALLY-PARALLEL ARCHITECTURE

A RTL model of a partially-parallel LDPC decoder using the 3L-HQC-LP matrix presented in Chapter 3, Structure and flexibility of LDPC codes (Section 3.2) has been designed using Verilog HDL. The hardware model has been designed for a ½ rate (3, 6) regular LDPC code, which is compliant with WLAN (IEEE Std. 802.11 n) and LTE application standards (see Table 5.1). This model is parameterized to support different code lengths and applications. Both the low-complexity algorithms presented in Section 4.2, namely SMP and MMS algorithms, have been used for FPGA implementation of the decoder. For both algorithms, 4-bit quantization of intrinsic messages has been used.

The top-level block diagram of the hardware model of the partially-parallel decoder is shown in Fig. 6.12.

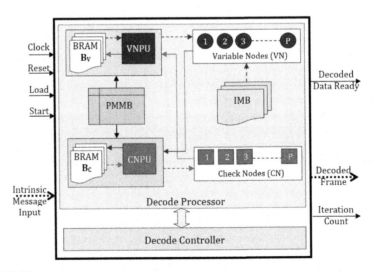

FIGURE 6.12

Top level block diagram of the partially-parallel LDPC decoder.

The decoder consists of two major blocks: Decode Controller (DC) and Decode Processor (DP). The DC is responsible for controlling the decoding process and responding to external control signals. It also organizes and sequences the input data and outputs the decoded data. The DP is responsible for the decoding process. It consists of Variable Node Processing Unit (VNPU), Check Node Processing Unit (CNPU), Variable Nodes (VN), Check Nodes (CN), Intrinsic Message Block (IMB), and the Permuted Matrix Memory Block (PMMB). Based on the required parallel nodes (P) for a particular application (see Section 3.2), the VN and CN blocks consist of chain of P variable nodes and P check nodes respectively. The Permuted random matrix information is stored in the form of Look-Up Tables (LUT) in PMMB. The VNPU and CNPU use these LUTs for accessing and storing messages at appropriate locations in the Block RAMs (BRAM). Appropriate HDL code is generated for PMMB, VN, and CN blocks using the automation technique described in Section 6.2.3.

To start with the decoding process, the VNPU first accesses the extrinsic messages from the BRAM (B_V) and passes it on to the VN. The VN processes this data along with the intrinsic message from IMB. The extrinsic messages generated by the variable nodes are pipelined to the CNPU for updating it in BRAM (B_C). The timing diagram in Fig. 6.13 illustrates the sequence of operations performed when VNP cycle is active. Each of the VNPU message processing cycle VN_J indicates P variable nodes operating in parallel. CM_J indicates P number of extrinsic messages from the variable nodes is updated in the BRAM (B_C). The VNP is active for J clock cycles till all the variable nodes are processed for the entire code length. The number of clock cycles J for the complete VNP operation is given by Eq. 6.3.

$$J = \frac{\text{Code length}}{\text{Parallel factor } (P)} \tag{6.3}$$

When VNP operation is complete, a similar message updating process is performed by CNPU next. The CNPU accesses the extrinsic messages from the BRAM (B_C) and passes it on to the CN. The extrinsic messages generated by the check nodes are pipelined to the VNPU for updating it in BRAM (B_V). The CN also outputs parity check information to the DC. The timing diagram in Fig. 6.14 illustrates the sequence of operations performed when CNP cycle is active. Each

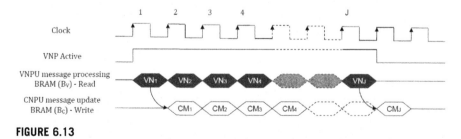

FIGURE 6.13

Timing diagram illustrating the operation of the variable node processing unit.

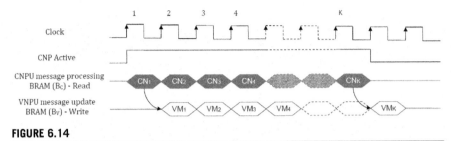

FIGURE 6.14

Timing diagram illustrating the operation of the check node processing unit.

of the CNPU message processing cycle CN_K indicates P check nodes operating in parallel. VM_K indicates P number of extrinsic messages from the check nodes is updated in the BRAM (B_V). The CNP is active for K clock cycles till all the check nodes are processed. The number of clock cycles K for the complete CNP operation is given by Eq. 6.4.

$$K = \frac{\text{Code rate} \times \text{Code length}}{\text{Parallel factor } (P)} \qquad (6.4)$$

This processing cycle of VNPU and CNPU completes a single decoding iteration of the decoder. The decoding process is stopped by DC when the maximum iteration count is reached or the parity check is satisfied. The decoder requires additional clock cycles to compensate the delays in VN and CN operations due to pipelined processing in each of the decoding iteration. The Latency (L) for the decoder is "6" clock cycles and is constant for any code lengths or parallel factors. The number of clocks per decoding iterations (CDI) for the partially-parallel LDPC decoder is computed as shown in Eq. 6.5 and Eq. 6.6.

$$CDI = J + K + L \qquad (6.5)$$

For example, CDI for a decoder using ½ rate 2304-bit LDPC code with parallel factor "16" is calculated as follows:

$$J = \left(\frac{2304}{16}\right) = 144$$

$$K = \left(\frac{\frac{1}{2} \times 2304}{16}\right) = 72$$

$$CDI = 144 + 72 + 6 = 222 \qquad (6.6)$$

The decoder architecture incorporates memory efficient design techniques to store extrinsic messages in BRAMs. The MMS algorithm itself provides huge saving in memory due to use of 2-bit messages compared to other conventional MS or SP based algorithms. In addition, the 2-bit messages from VN and CN are concatenated to form a single word of 2P-bit message by the node processing units (VNPU and CNPU) and are stored in a single address location in BRAM.

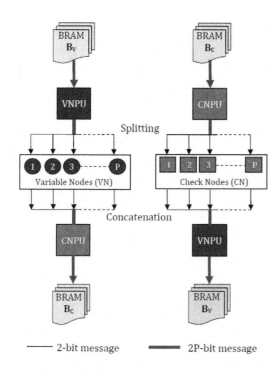

FIGURE 6.15

Illustration of memory efficient design of the decoder.

While retrieving the messages, a single word of 2P-bit is read from BRAM and is split into 2-bit messages by the node processing unites and passed to the respective VN or CN for processing. Hence each cycle of CNPU or VNPU operation involves read/write of 2P-bit word in BRAM. This technique efficiently utilizes BRAMs and also significantly reduces the complexity of the VNPU and CNPU blocks for message exchange. The message concatenation and splitting operation of the decoder is illustrated in Fig. 6.15.

The VNPU and CNPU consist of similar finite state machines (FSM) that operate in parallel for the entire decode operation. When VNPU is active, the CNPU is in the update state. Similarly, when CNPU is active, the VNPU is in the update state. A simple illustration of the implemented FSMs is shown in Fig. 6.16. The operation of various blocks of the decoder and its states are also shown in Table 6.5.

6.3.3 PERFORMANCE ANALYSIS

As an example, the FPGA implementation results for ½ rate (3, 6) 2304-bit LDPC code (for LTE) using MMS algorithm have been analyzed. The BER, FER

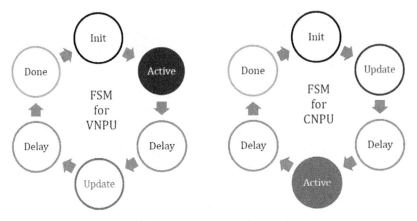

FIGURE 6.16

Illustration of FSMs for VNPU and CNPU of the decoder.

Table 6.5 Different States of Operation for Various Blocks of the Decoder

	VNPU	CNPU	VN	CN	BRAM (B_V)	BRAM (B_C)
"*J*" clock cycles	Active	Update	Enabled	Disabled	Read	Write
"*K*" clock cycles	Update	Active	Disabled	Enabled	Write	Read

performance and average iterations for the partially-parallel decoder are shown in Figs. 6.17, 6.18, and 6.19 respectively. From Figs. 6.17 and 6.18, it is noted that the BER and FER performances of the decoder implemented on FPGA closely follow the performances of the software-based decoder model with a slight loss of about 0.15 dB. As stated earlier (see Section 6.3.2), this loss is due to the approximation of mapping LUTs for MMS decoding algorithm. However, the average iterations for the software model and the FPGA-based decoder are same as shown in Fig. 6.19.

The throughput (*T*) of the partially-parallel decoder is computed using Eq. 6.2. The number of parallel nodes (*P*) in the decoder implemented on FPGA is 16 and code length is ½ rate 2304-bit. From Eq. 6.6, the number of clocks per decoding iteration (CDI) for this decoder is 222. Therefore, with a maximum operating frequency of 162 MHz (from postplacement and routing report), the average throughput of the FPGA-based LDPC decoder at 3.75 dB E_b/N_o is 115 Mbps.

The benefits of partially-parallel decoders over fully-parallel architectures can be seen from the comparative data presented in Table 6.6. The partially-parallel architecture offers flexibility in implementing decoders with various code lengths. This flexibility is predominantly due to the use of the HQC-LP-based LDPC matrix (see Section 3.2). A key feature of the design is that the logic resources

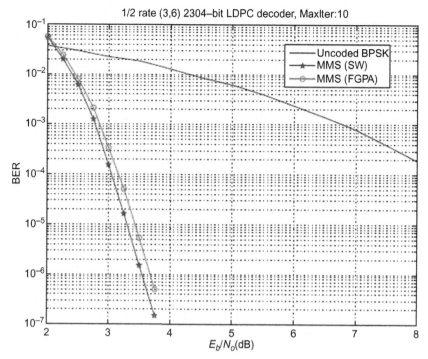

FIGURE 6.17

BER performance of the partially-parallel MMS decoder implemented on FPGA.

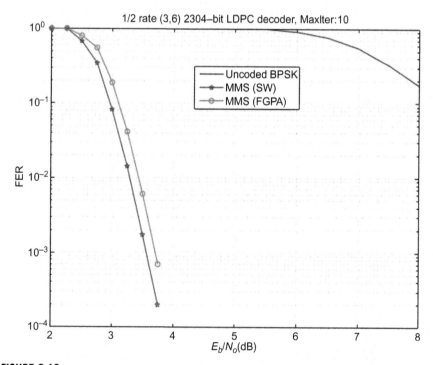

FIGURE 6.18

FER performance of the partially-parallel MMS decoder implemented on FPGA.

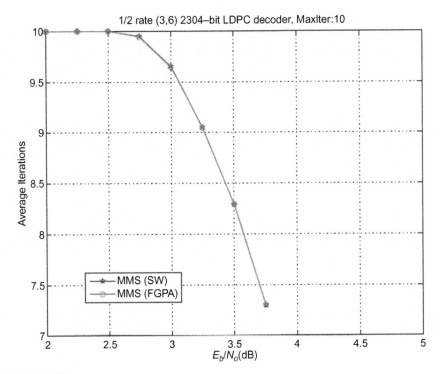

FIGURE 6.19

Average decoding iterations for the partially-parallel MMS decoder implemented on FPGA.

Table 6.6 FPGA Implementation Results for Partially-Parallel and Fully-Parallel Decoders

	Partially-Parallel			**Fully-Parallel**
Algorithm	MMS (C1)			
LDPC code	576	1152	2304	1152
Parallel nodes (variable and check)	16 VN and 16 CN			1152 VN and 576 CN
Slices	1,137 (6%)			10,823 (62%)
LUTs	3,522 (5%)			39,024 (56%)
Registers	847 (1.2%)			15,107 (22%)
BRAMs	29 (10%)			None
Clocks per decoding iteration (CDI)	60	114	222	1
Clock frequency (MHz)	162			138
Avg. Throughput at BER of 10^{-6} (Mbps)	142	124	104	11700
Total power (W)	0.990			1.13
FPGA device	Xilinx Virtex 5 (XC5VLX110T-3FF1136)			

(LUTs and Registers) remain constant irrespective of the code length. These account for less than 10% of the logic resources required by a fully-parallel implementation. The number of block RAMs also remain unaltered for various code lengths. There is a slight saving in power consumption compared to a fully-parallel design. However, a fully-parallel decoder can potentially achieve a massive throughput compared to a partially-parallel decoder. The throughput of the latter can be improved by increasing the number of parallel nodes in the partially-parallel architecture. The implementation results for such decoders are discussed later in this chapter.

A comparison of the hardware requirements and performance of the partially-parallel LDPC decoder with those of other partially-parallel decoders are given in Table 6.7. Among many other partially-parallel decoder architectures [6,8,35−38] reported to date, the ones that have configuration close to the presented decoder are compared in this table. Decoders with similarities such as LDPC code, application, and FPGA device are chosen for comparison. It is observed that the presented architecture requires substantially less hardware resources compared to other LTE application based partially-parallel decoders. With significantly reduced memory (bits) and BRAM requirement, the presented LDPC decoder also achieves comparable throughput to those of the other decoders [39] and [40]. Table 6.7 clearly illustrates the area and memory efficient design of the presented partially-parallel decoder architecture. This is achievable mainly due to the use of the reduced complexity MMS algorithm (see Section 4.2.2) with node optimization and concatenation of messages in BRAM. The architecture of the decoder is easily scalable to achieve higher throughput by increasing the number of parallel nodes (P).

Table 6.7 Comparison with Other FPGA-Based Partially-Parallel LDPC Decoders

	Presented	[39]	[41]	[40]	[42]	[7]
Application	LTE/WiMAX					
Slices	3806	6,568	20,746	NA	NA	NA
LUTs/ALUTs	6.796	11,028	33,226	19,000	27,850	17,259
Registers	598	6330	32,619	10,000	9806	6598
BRAMs	29	100	75	92	NA	NA
Total memory (bits)	20,736	60,288	NA	NA	100,552	271,104
Clock frequency (MHz)	84	110	192.4	160	100	155
Avg. Throughput (Mbps)	55	61	NA	10.4	154	232.5
FPGA device	Virtex 2	Virtex 2	Virtex 4	Virtex 5	Stratix 2	Stratix 2

Table 6.8 Comparison with ASIC-Based Partially-Parallel LDPC Decoders

	Presented	**[47]**	**[43]**	**[44]**	**[48]**	**[49]**
Application	LTE / WiMAX					
Implementation	Virtex 5 FPGA	ASIC				
IC technology (nm)	65	90	90	130	130	180
Clock frequency (MHz)	162	150	400	145	83.3	100
Throughput (Mbps)	104	105	133	302	60	68
Total power (mW)	990	264	901	170	52	165

However, this leads to an increase in hardware resources, particularly the BRAM. A detailed exploration of design space and analysis of performance for the presented LDPC decoder based on various parameters is presented in the next section.

The presented partially-parallel decoder is also compared for speed and power against other ASIC-based decoders reported in the literature. The comparison data is shown in Table 6.8. The throughput performance of the presented decoder closely matches those of the ASIC-based decoders, except [43] and [44]. However the power requirement [45,46] of the presented FPGA decoder is far greater than ASIC decoders.

6.4 DESIGN SPACE EXPLORATION

Further detailed analysis of decoder performances based on various parameters (BER, speed, area and power) and configurations (applications, LDPC codes and parallel factor) is presented in this section for both SMP and MMS algorithms. Fully-parallel and partially-parallel architectures suitable for WLAN and LTE applications using ½ rate (3,6) regular LDPC codes are considered for design space exploration. The performance analysis results for fully-parallel and partially-parallel decoders are provided in Table 6.9 and Table 6.10 respectively.

As an example, graphical results on performance are illustrated in the next section for a partially-parallel decoder using the MMS algorithm. The FPGA implementation results are obtained from a Xilinx Virtex5 (XC5VLX110 T) device. The logic and memory resources available on this FPGA are listed in Table 6.11.

6.4.1 DECODING PERFORMANCE

The BER performances of partially-parallel MMS decoders for WLAN and LTE applications are shown in Figs. 6.20 and 6.21 respectively. In these figures, the label "SW" indicates results obtained from the software simulation model and the

Table 6.9 Comparison of Design Space and Performance of the Presented Fully-Parallel LDPC Decoders

		Fully-Parallel Architecture							
Algorithm		**MMS**				**SMP**			
Application		**WLAN**		**LTE**		**WLAN**		**LTE**	
LDPC code		648	1296	576	1152	648	1296	576	1152
Node interconnections (I)		7776	15,552	6912	13,824	3888	7776	3456	6912
Slices		7048	12,968	6218	11,984	4455	8490	3763	7448
LUTs		21,952	43,823	19,523	38,962	14,177	28,272	12,611	25,139
Registers		8489	16,913	7553	15,041	5897	11729	5249	10,433
Clock (MHz)		113	66	116	82	117	74	140	92
At BER of 10^{-6}	E_b/N_o (dB)	4.3	3.8	4.3	3.9	7.75	7.25	7.8	6.65
	Avg. Iterations	5.6	7.1	5.5	6.6	2.6	3.4	2.5	3.8
	Avg. Throughput (Gbps)	6.5	6	6.1	7.1	14.6	14.1	16.1	13.9
Static power (W)		0.984	1.116	0.983	1.034	0.982	0.985	0.982	0.985
Dynamic power (W)		0.057	0.104	0.048	0.096	0.043	0.069	0.035	0.069
Total power (W)		1.041	1.22	1.032	1.13	1.025	1.055	1.016	1.055

Table 6.10 Comparison of Design Space and Performance of the Presented Partially-Parallel LDPC Decoders

	Partially-Parallel Architecture															
Algorithm	**MMS**								**SMP**							
Application	**WLAN**				**LTE**				**WLAN**				**LTE**			
LDPC code	**1944**				**2304**				**1944**				**2304**			
Parallel factor (P_f)	1	3	6	9	1	3	6	9	1	3	6	9	1	3	6	9
Parallel nodes (P)	18	54	108	162	16	48	96	144	18	54	108	162	16	48	96	144
Slices	1277	3336	6327	9490	1137	3141	5583	8430	843	1895	3564	5093	734	1610	3216	4682
LUTs	3897	10,701	20,818	31,227	3522	9547	18,542	27,558	2275	5793	11,495	16,747	2075	5122	9829	14,956
Registers	945	2270	4484	6726	847	2024	3992	5961	749	1676	3297	4919	671	1498	2936	4378
BRAM (18 K)	29	87	174	261	29	87	160	232	17	51	102	153	17	51	100	136
Total memory (bits)	17,496				20,736				13,608				16,128			
Clock (MHz)	150	130	124	108	162	144	126	114	176	142	128	124	175	136	122	120
Clocks per decoding iteration (CDI)	168	60	33	24	222	78	42	30	168	60	33	24	222	78	42	30
At BER of 10^{-6} — E_b/N_o (dB)	3.7				3.6				7.1				7			
At BER of 10^{-6} — Avg. Iterations	7.5				8.0				3.6				3.8			
Avg. Throughput (Mbps)	116	280	486	582	104	266	432	548	282	639	1047	1395	239	528	880	1212
Static power (W)	0.979	0.981	0.984	0.987	0.979	0.981	0.983	0.985	0.979	0.980	0.982	0.984	0.978	0.979	0.982	0.983
Dynamic power (W)	0.011	0.031	0.058	0.078	0.011	0.029	0.051	0.069	0.011	0.020	0.041	0.055	0.005	0.017	0.038	0.048
Total power (W)	0.990	1.012	1.042	1.065	0.990	1.010	1.034	1.055	0.990	1.000	1.023	1.039	0.984	0.996	1.019	1.031

Table 6.11 Resources Available in a Xilinx Virtex5 FPGA

Resources	Available
Slices	17,280
LUTs	69,120
Registers	69,120
Block RAMs (18 K)	296
Total Memory (KB)	5328
IO ports	640
Device	XC5VLX110T-3FF1136

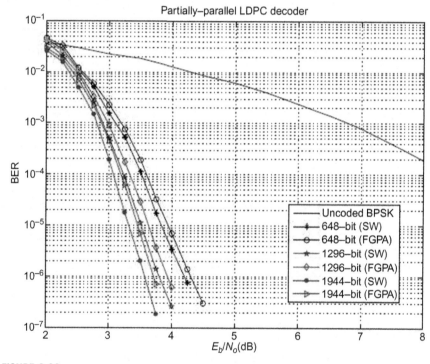

FIGURE 6.20

BER performance of the partially-parallel MMS decoder for WLAN.

label "FPGA" indicates results obtained from FPGA implementation. It is observed that the performance of the decoders is confined within a range of 3.5 dB to 4.5 dB E_b/N_o to achieve a BER of 10^{-6}. A loss of about 0.05 dB is noticed in cases of the decoder implemented on FPGA compared to its software

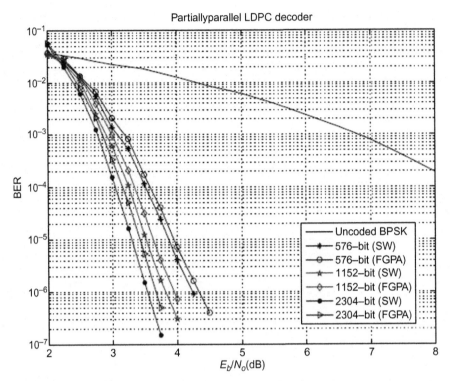

FIGURE 6.21

BER performance of the partially-parallel MMS decoder for LTE.

models. This is due to practical implementation errors such as quantization and fixed-point operations. However, the average iterations of the FPGA-based partially-parallel MMS decoders closely follow those of the software models, as shown in Figs. 6.22 and 6.23 for WLAN and LTE applications respectively.

6.4.2 HARDWARE PERFORMANCE

Figs. 6.24 and 6.25 show the estimated average throughput of the partially-parallel MMS decoders for WLAN and LTE applications respectively over a range of parallelism factors (P_f). The number of parallel nodes (check nodes and variable nodes) used in the decoder is indicated by the multiple of the factor P_f, as shown in Eq. 6.7. These figures reveal that as P_f increases, there is a corresponding increase in throughput, although operating frequency of the decoder is proportionally reduced. The improvement in the throughput is due to the increased number of parallel nodes in the decoder.

$$\text{Number of parallel nodes } (p) = P_f \times P_A \qquad (6.7)$$

where P_f is a positive integer; $P_A = 18$ for WLAN and $P_A = 16$ for LTE.

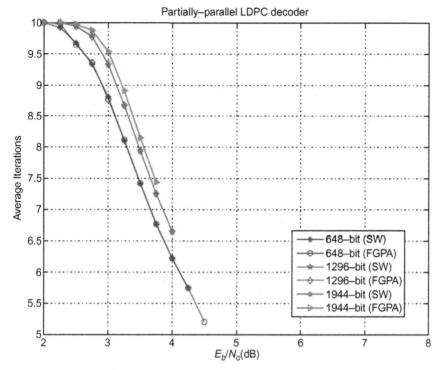

FIGURE 6.22

Average iterations of the partially-parallel MMS decoder for WLAN.

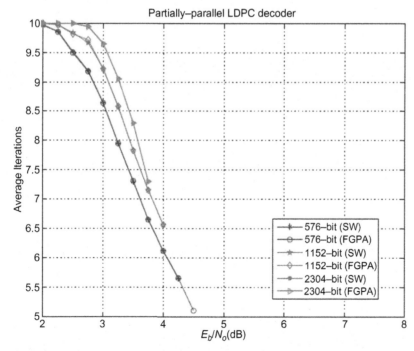

FIGURE 6.23

Average iterations of the partially-parallel MMS decoder for LTE.

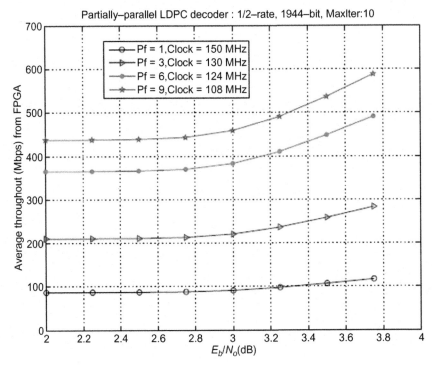

FIGURE 6.24

Average throughput of the partially-parallel MMS decoder for WLAN.

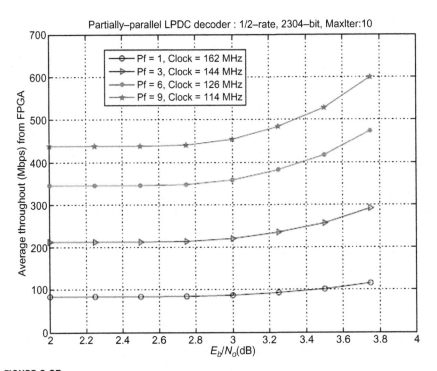

FIGURE 6.25

Average throughput of the partially-parallel MMS decoder for LTE.

The FPGA logic utilization statistics for the partially-parallel MMS decoder is shown in Fig. 6.26. It is evident that as the parallel factor (P_f) increases, the logic requirements for the decoder also increases. The overall logic requirements for the decoder are very close for WLAN and LTE applications for a given P_f.

The total power consumption of the decoder (including static and dynamic power) is analyzed for a range of P_f. The results are shown in Fig. 6.27. It is observed that the dynamic power consumed by the decoder is insignificant compared to the static power. The total power requirement for the decoder remains more or less constant for both WLAN and LTE applications.

Fig. 6.28 shows a gross comparison of the decoders in terms of memory requirement (BRAMs), frequency of operation (f_{max}), clocks per decoding iterations, and throughput. It is noted that as the number of parallel nodes increase, there is a corresponding increase in BRAMs required to store more parallel messages. This increase in parallel nodes result in a reduced number of clock cycles required for completing a decoding iteration. Due to the increased complexity introduced by the increased number of parallel nodes, the maximum operating

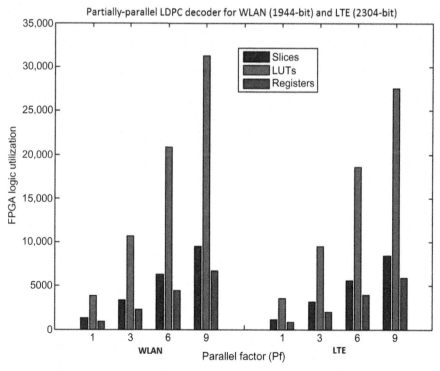

FIGURE 6.26

FPGA logic utilization of the decoder for WLAN and LTE.

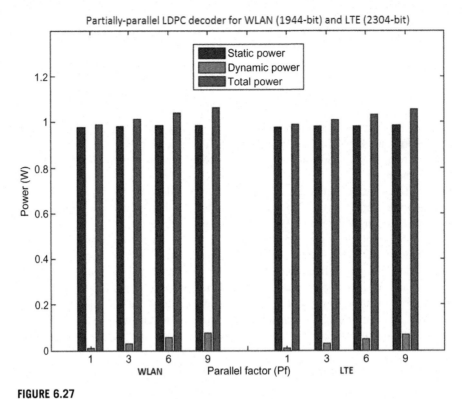

FIGURE 6.27

Power dissipation of the FPGA-based decoders for WLAN and LTE.

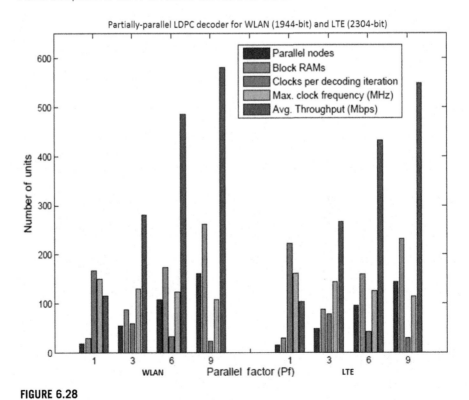

FIGURE 6.28

Performance comparison of the decoder for WLAN and LTE.

frequency is reduced. However, the average throughput of the decoder improves significantly due to the use of a larger number of nodes in parallel.

6.5 SUMMARY

The product of area, power, and delay of a decoder indicates the overall performance and efficiency of the implemented architecture [50]. The smaller the resultant product, the better is the efficiency of the decoder. The product is calculated in Eq. 6.8.

$$E_{APD} = \text{Area} \times \text{Power} \times \text{Delay} \tag{6.8}$$

where Area = Number of FPGA slices and number of memory bits utilized by the decoder; Power = Total power consumed by the decoder (static + dynamic); Delay = Maximum operating delay of the decoder $(1/f_{max})$.

Fig. 6.29 shows the E_{APD} factor of partially-parallel decoders for both SMP and MMS algorithms for a range of parallel factors (P_f). It is observed that E_{APD}

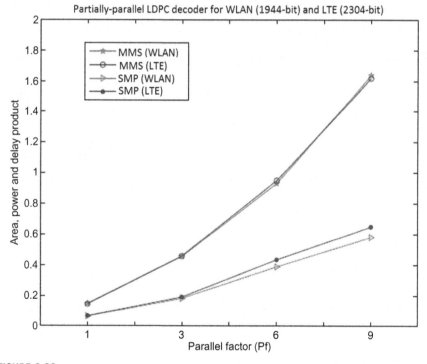

FIGURE 6.29

Overall performance characteristics of the LDPC decoder.

for the SMP-based decoders is much less than that of the MMS-based decoders. It can also be noted that with increasing P_f, the E_{APD} factor shows an increasing trend for both WLAN and LTE, which is expected.

REFERENCES

[1] V.A. Chandrasetty, S.M. Aziz, FPGA implementation of a LDPC decoder using a reduced complexity message passing algorithm, J. Netw. Acad. Publ. 6 (1) (2011) 36–45.

[2] NIAGARA Flow - High Performance LDPC decoder for 10 Gbps Ethernet. [cited Nov 2009]. Available from: www.widesail.com.

[3] FRAZER flow - LDPC decoder for 100 Gbps Optical Comms. [cited Nov 2009]. Available from: www.widesail.com.

[4] H. Zhiyong, R. Sebastien, F. Paul. FPGA implementation of LDPC decoders based on joint row-column decoding algorithm. in: IEEE International Symposium on Circuits and Systems, New Orleans, LA, 2007.

[5] K. Shimizu, et al. Partially-parallel LDPC decoder based on high-efficiency message-passing algorithm. in: IEEE International Conference on Computer Design: VLSI in Computers and Processors, 2005.

[6] C. Zhiqiang, W. Zhongfeng, L. Youjian, High-throughput layered LDPC decoding architecture, IEEE Trans. Very Large Scale Integr. (VLSI) Syst. 17 (4) (2009) 582–587.

[7] H. Ding, et al. Design and implementation for high speed LDPC decoder with layered decoding. in: WRI International Conference on Communications and Mobile Computing, Yunnan, 2009.

[8] D. Yongmei, C. Ning, Y. Zhiyuan, Memory efficient decoder architectures for quasi-cyclic LDPC codes, IEEE Trans. Circuits Syst. I Regul. Pap. 55 (9) (2008) 2898–2911.

[9] Z. Chuan, et al., Flexible LDPC decoder design for multigigabit-per-second applications, IEEE Trans. Circuits Syst. I Regul. Pap. 57 (1) (2010) 116–124.

[10] M.D. Ciletti, Advanced Digital Design with the Verilog HDL, second ed., Prentice Hall, 2010.

[11] S.M. Aziz, M.D. Pham, Implementation of low density parity check decoders using a new high level design methodology, J. Comput. Acad. Publ. 5 (1) (2010) 81–90.

[12] R. Zarubica, S.G. Wilson, E. Hall. Multi-Gbps FPGA-based low density parity check (LDPC) decoder design. in: IEEE Global Telecommunications Conference. 2007. Washington, DC.

[13] V.A. Chandrasetty, *VLSI Design: A practical guide for FPGA and ASIC implementations*. Springer Briefs in Electrical and Computer Engineering, Vol. 1, Springer, 2011, p. 120.

[14] FPGA Basics. [cited October 2017]. Available from: <https://www.xilinx.com/products/silicon-devices/fpga/what-is-an-fpga.html>.

[15] FPGA Architectures. [cited May 2017]. Available from: <http://www.springer.com/cda/content/document/cda_downloaddocument/9781461435938-c2.pdf>.

[16] Introduction to FPGA Technology: Top Five Benefits. [cited May 2017]. Available from: <http://zone.ni.com/devzone/cda/tut/p/id/6984>.

[17] FPGA vs ASIC. [cited October 2017]. Available from: <https://www.xilinx.com/fpga/asic.htm>.

[18] Z. Kai, H. Xinming, W. Zhongfeng, A high-throughput LDPC decoder architecture with rate compatibility, IEEE Trans. Circuits Syst. I Regul. Pap. 58 (4) (2011) 839−847.

[19] X. Qian, et al. A high parallel macro block level layered LDPC decoding architecture based on dedicated matrix reordering. in: IEEE Workshop on Signal Processing Systems, Beirut, Lebanon, 2011.

[20] M. Weiner, B. Nikolic, Z. Zhengya. LDPC decoder architecture for high-data rate personal-area networks. in: IEEE International Symposium on Circuits and Systems, Rio de Janerio, Brazil, 2011.

[21] S. Sun, H. Qi. A pipelining hardware implementation of H.264 based on FPGA. in: International Conference on Intelligent Computation Technology and Automation, Changsha, 2010.

[22] K. Bongjin, P. In-Cheol. QC-LDPC decoding architecture based on stride scheduling. in: IEEE International Symposium on Circuits and Systems, Rio de Janeiro, Brazil, 2011.

[23] D. Divsalar, S. Dolinar, C. Jones. Low-rate LDPC codes with simple protograph structure. in: International Symposium on Information Theory, Adelaide, Australia, 2005.

[24] P. Duc Minh, S.M. Aziz. FPGA architecture for object extraction in Wireless Multimedia Sensor Network. in: Seventh International Conference on Intelligent Sensors, Sensor Networks and Information Processing, Adelaide, SA, 2011.

[25] M.N. Sakib, et al. Low complexity soft decision circuit for LDPC decoders. in: Conference on Lasers and Electro-Optics, San Jose, California, 2011.

[26] D.J. Costello Jr, et al. A comparison between LDPC block and convolutional codes. in: Information Theory and Application Workshop, 2006.

[27] R.M. Tanner, et al., LDPC block and convolutional codes based on circulant matrices, IEEE Trans. Inf. Theory 50 (12) (2004) 2966−2984.

[28] T.B. Iliev, et al., Application and evaluation of the LDPC codes for the next generation communication systems, Automation and Industrial Electronics Novel Algorithms and Techniques In Telecommunications (2008) 532−536.

[29] V.A. Chandrasetty, S.R. Laddha. A novel dual processing architecture for implementation of motion estimation unit of H.264 AVC on FPGA. in: IEEE Symposium on Industrial Electronics & Applications, Kuala Lumpur, 2009.

[30] V.A. Chandrasetty, S.M. Aziz. FPGA implementation of high performance LDPC decoder using modified 2-bit min-sum algorithm. in: 2nd International Conference on Computer Research and Development, Kuala Lumpur, 2010.

[31] S.S. Tehrani, S. Mannor, W.J. Gross. An Area-Efficient FPGA-Based Architecture for Fully-Parallel Stochastic LDPC Decoding. in: IEEE Workshop on Signal Processing Systems, Shanghai, China, 2007.

[32] S. Sharifi Tehrani, S. Mannor, W.J. Gross, Fully parallel stochastic LDPC decoders, IEEE Trans. Signal Process. 56 (11) (2008) 5692−5703.

[33] A. Rapley, et al. Stochastic iterative decoding on factor graphs. in: 3rd International Symposium on Turbo Codes and Related Topics, 2003.

[34] W.J. Gross, V.C. Gaudet, A. Milner. Stochastic implementation of LDPC decoders. in: 39th Asilomar Conference on Signals, Systems and Computers, Pacific Grove, CA, 2005.

[35] C. Zhiqiang, W. Zhongfeng. A 170 Mbps (8176, 7156) quasi-cyclic LDPC decoder implementation with FPGA. in: IEEE International Symposium on Circuits and Systems, Island of Kos, Greece, 2006.

[36] X. Zhang, F. Cai. Partial-parallel decoder architecture for quasi-cyclic non-binary LDPC codes. in: International Conference on Acoustics Speech and Signal Processing, 2010.

[37] Z. Xinmiao, C. Fang, Efficient partial-parallel decoder architecture for quasi-cyclic nonbinary LDPC codes, IEEE Trans. Circuits Syst. I Regul. Pap. 58 (2) (2011) 402−414.

[38] W. Wang, et al. A 223 Mbps FPGA implementation of (10240, 5120) irregular structured low density parity check decoder. in: IEEE Vehicular Technology Conference, Singapore, 2008.

[39] K.K. Gunnam, et al. VLSI architectures for layered decoding for irregular LDPC codes of WiMax. in: IEEE International Conference on Communications, Glasgow, 2007.

[40] F. Charot, et al. A new powerful scalable generic multi-standard LDPC decoder architecture. in: 16th International Symposium on Field-Programmable Custom Computing Machines, Palo Alto, CA, 2008.

[41] C. Xiaoheng, L. Shu, V. Akella, QSN: a simple circular-shift network for reconfigurable quasi-cyclic LDPC decoders, IEEE Trans. Circuits Syst. II Express Briefs 57 (10) (2010) 782−786.

[42] QC-LDPC Decoder IP Core. [cited April 2009]. Available from: <http://unicore.co.ua/index.php?page = products&hl = en>.

[43] W. Yu-Luen, et al. A low-complexity LDPC decoder architecture for WiMAX applications. in: International Symposium on VLSI Design, Automation and Test. Hsinchu, 2011.

[44] B. Dan, et al. A 4.32 mm 2 170 mW LDPC decoder in 0.13 υm CMOS for WiMax/Wi-Fi applications. in: 16th Asia and South Pacific Design Automation Conference, Yokohama, 2011.

[45] I. Kuon, J. Rose, Measuring the gap between FPGAs and ASICs, IEEE Trans. Comput. Aided Design Integr. Circuits Syst. 26 (2) (2007) 203−215.

[46] A. Amara, F. Amiel, T. Ea, FPGA vs. ASIC for low power applications, Microelectronics J. 37 (8) (2006) 669−677.

[47] L. Chih-Hao, et al., An LDPC decoder chip based on self-routing network for IEEE 802.16e applications, IEEE J. Solid-State Circuits 43 (3) (2008) 684−694.

[48] X.-Y. Shih, C.-Z.Z., C.-H. Lin, A.-Y. (Andy) Wu. A 52-mW 8.29mm2 19-mode LDPC Decoder Chip for Mobile WiMAX Applications. in: Asia and South Pacific Design Automation Conference, Yokohama, 2009.

[49] K. Tzu-Chieh, A.N. Willson. A flexible decoder IC for WiMAX QC-LDPC codes. in: IEEE Custom Integrated Circuits Conference, San Jose, USA, 2008.

[50] E. Yeo, V. Anantharam, Capacity approaching codes, iterative decoding architectures, and their applications, IEEE Commun. Mag. 41 (8) (2003) 132−140.

LDPC decoders in multimedia communication

7.1 IMAGE COMMUNICATION USING LDPC CODES

Performance evaluation has been carried out by transmitting images over an AWGN channel and reconstructing the received images using the partially-parallel MMS decoder in MATLAB environment. Bitmap (BMP) and JPEG files with samples of color images have been used in the simulations. In [1,2], LDPC codes were used to protect uncompressed grayscale images from errors. For better protection of baseline JPEG images, an Unequal Error Protection (UEP) [3] scheme using LDPC codes and Reed-Solomon (RS) codes was presented in [4]. Performance evaluations of a hybrid combination of RS and LDPC codes in [5,6] show increased reliability in transmission of multimedia content.

The evaluation scheme for BMP and JPEG images is illustrated in Fig. 7.1. For BMP images, the evaluation technique is simple as shown in Fig. 7.1A. However, for JPEG images the encoding scheme requires an additional level of security due to the presence of headers in JPEG format. A hybrid encoding technique presented in [6,7] is incorporated in the communication system to ensure reliable transmission of JPEG image data as shown in Fig. 7.1B. This technique uses Reed-Solomon (RS) codes for encoding header and tail sections of the JPEG image [8]. The RS encoded headers along with the rest of the JPEG data is then encoded using LDPC codes [5,9]. The RS-LDPC encoded data is transmitted over an AWGN channel, where the data is deliberately subjected to some errors. The received erroneous data is first decoded using the LDPC decoder. Then the header/tail sections of the JPEG image are decoded using a RS decoder for image reconstruction. The loss in compression rate of the image due to introduction of RS encoding is negligible when compared to the data integrity of multimedia content achieved using such encoding techniques [4].

7.2 PERFORMANCE ANALYSIS

The quality of the reconstructed images has been compared against the original image under various BER conditions and different code lengths of the partially-parallel LDPC decoder in [10]. Color images of size 512×512 pixels in BMP

Resource Efficient LDPC Decoders. DOI: https://doi.org/10.1016/B978-0-12-811255-7.00007-1

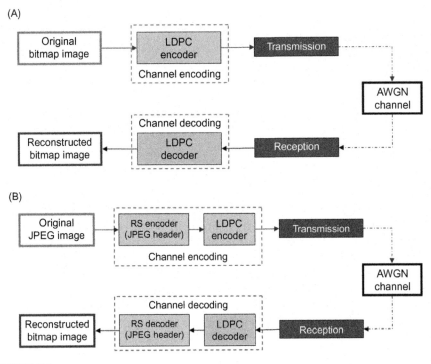

FIGURE 7.1

Performance evaluation of LDPC decoders for multimedia communication.
(A) Performance evaluation scheme for LDPC decoder using Bitmap images.
(B) Performance evaluation scheme for LDPC decoder using JPEG images.

and JPEG2000 [11] formats have been used for simulations. The partially-parallel LDPC decoder that complies with the LTE application standard [12] has been chosen for the analysis (see Section 6.3.2).

The quality of the reconstructed images at various BER conditions and code lengths have also been analyzed by calculating the Peak Signal to Noise Ratio (PSNR) with respect to the original image [1]. The PSNR is calculated using Eqs. 7.1 and 7.2 [13].

$$PSNR \text{ (dB)} = 10 \times \log\left(\frac{P_{max}^2}{MSE}\right) \quad (7.1)$$

$$MSE = \sum_{i=1}^{x}\sum_{j=1}^{y}\frac{\left(|A_{ij}-B_{ij}|\right)^2}{x \times y} \quad (7.2)$$

where, MSE: Mean-Square Error; P_{max}: Maximum value of a pixel in the image; A: Pixel value of original image; B: Pixel value of reconstructed image; x: Height of the image in pixels; y: Width of the image in pixels

7.2.1 **QUALITY OF THE RECONSTRUCTED BMP IMAGES**

Visual comparison of the quality of the original and reconstructed BMP images at various BER conditions is shown in Table 7.1. It is observed that the images reconstructed at a BER of 10^{-6} have negligible difference compared to the original images. However, at a BER of 10^{-4} there is a slight noise in the images [14]. When the BER is further increased to 10^{-2}, the reconstructed images show significant effect of salt-pepper noise [15]. As expected, the quality of the reconstructed image improves as the bit errors in the decoded data reduce.

The BER versus PSNR plot for the reconstructed BMP images is shown in Fig. 7.2. Clearly, the PSNR range is higher for images transmitted at lower bit error condition. For example, PSNR >70 dB at a BER of 10^{-6} and PSNR <50 dB at a BER of 10^{-2}. The effect of the BER performance of the decoder on the reconstructed images can be verified by visually inspecting the quality of the reconstructed images presented in Table 7.1.

The quality of the reconstructed BMP images has also been evaluated for different code lengths of the partially-parallel MMS LDPC decoder. The code lengths versus PSNR plot is shown in Fig. 7.3. It is observed that the PSNR for the reconstructed images are almost constant for shorter code length (576-bit) and it significantly improves as the code length increases.

7.2.2 **QUALITY OF THE RECONSTRUCTED JPEG IMAGES**

A visual comparison of the quality of JPEG images at various BER conditions is shown in Table 7.2. The images reconstructed at a BER of 10^{-6} have negligible difference compared to the original images [16]. However, at a BER of 10^{-5} there is a slight drift in the luminance component of the images. When BER is further increased to 10^{-4} the reconstructed images are significantly distorted both in luminance and in spatial component. As expected, the quality of the reconstructed images deteriorates as the bit errors in the decoded data increases.

The BER versus PSNR plot for the reconstructed JPEG images is shown in Fig. 7.4. It is observed that PSNR >70 dB at a BER of 10^{-6} and PSNR <50 dB at a BER of 10^{-4}. The code lengths versus PSNR plot is shown in Fig. 7.5. This figure illustrates that using decoders with larger code lengths results in higher PSNR range and eventually leads to better quality of reconstructed images.

7.2.3 **RECONSTRUCTED JPEG IMAGES FOR VARIOUS DECODERS**

The quality of the reconstructed JPEG images for the proposed partially-parallel 3L-HQC decoder has been compared and analyzed against PEG- and QC-based LDPC decoders. All the decoders were simulated using the MMS algorithm, ½ rate (3, 6) 2304-bit LDPC code and a maximum of 10 iterations. The BER performance of the decoders simulated under similar conditions is shown in Fig. 7.6 for

Table 7.1 Comparison of the Quality of the Original and Reconstructed BMP Images

	Original Image	Reconstructed Image at BER of		
		10^{-6}	10^{-4}	10^{-2}
(1) Barbara				
(2) Parrot				
(3) Eiffel				

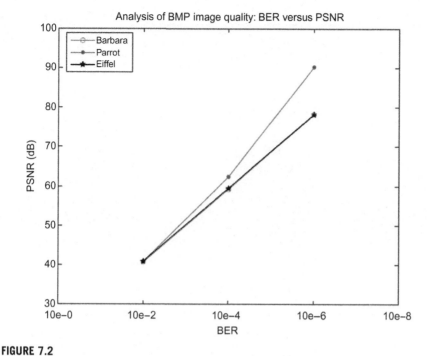

FIGURE 7.2

BER versus PSNR for the reconstructed BMP images.

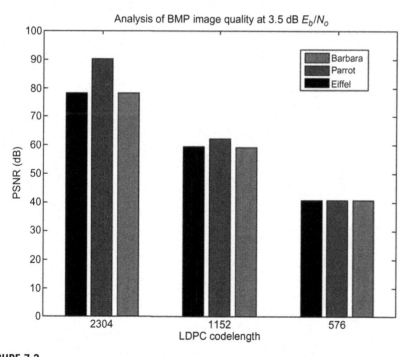

FIGURE 7.3

LDPC code length versus PSNR for the reconstructed BMP images.

Table 7.2 Comparison of the Quality of the Original and Reconstructed JPEG Images

	Original Image	Reconstructed Image at BER of		
		10^{-6}	10^{-5}	10^{-4}
(1) Lena				
(2) Baboon				
(3) Golden Gate				

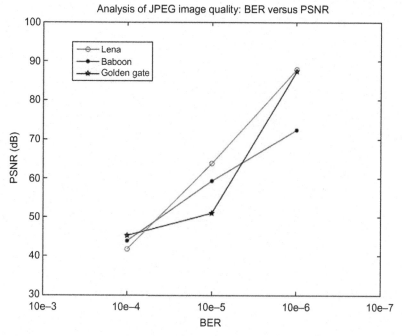

FIGURE 7.4

BER versus PSNR for the reconstructed JPEG images.

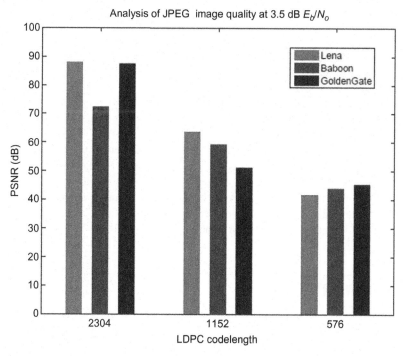

FIGURE 7.5

LDPC code length versus PSNR for the reconstructed JPEG images.

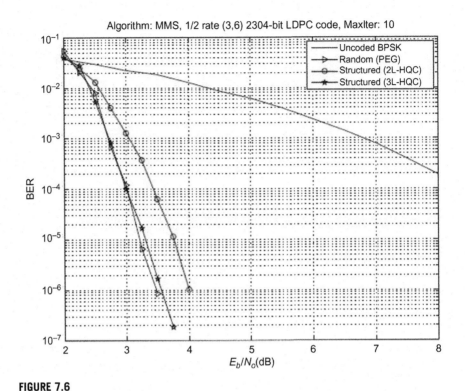

FIGURE 7.6

BER performance of various LDPC decoders using MMS algorithm.

reference. It is observed that at an E_b/N_o of 3.5 dB, the BER performance of the proposed decoder and the PEG-based decoder are much better than that of the 2L-HQC-based decoder.

The BER versus PSNR plots for these LDPC decoders using a JPEG image sample (Lena) are shown in Fig. 7.7. The PSNR values for PEG and the proposed decoder are similar over the BER range of 10^{-4} to 10^{-6}. However, 2L-HQC-based LDPC matrix has comparatively lower PSNR values over the same BER range. The same aspect has been verified by analyzing the visual quality of the reconstructed images as presented in Table 7.3.

7.3 SUMMARY

This chapter presented an evaluation scheme to analyze the performance of the partially-parallel MMS LDPC decoder for multimedia communication.

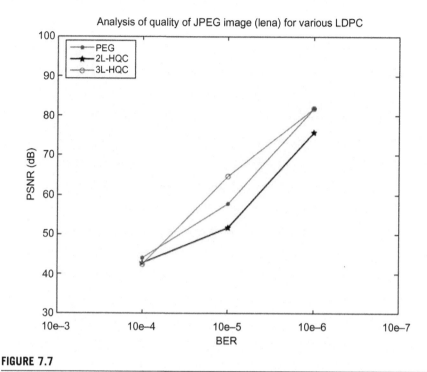

FIGURE 7.7

BER versus PSNR for Lena image sample using various LDPC decoders.

Simulations were carried out to assess the quality of BMP and JPEG images when transmitted over an AWGN channel. It is shown that the quality of the reconstructed images is high under low BER conditions and improves with longer code lengths of the LDPC decoder. From the results, it is also observed that the impact of BER on PSNR and image quality is significant for JPEG images compared to BMP images. This phenomenon can be observed by comparing the PSNR values for the BMP (see Fig. 7.2) and JPEG images (see Fig. 7.4) at a BER of 10^{-4}. The PSNR is about 60 dB for BMP images, whereas it is around 40 to 50 dB for JPEG images. This indicates that the quality of the reconstructed BMP images is better compared to the JPEG images at a given BER. Also, it is verified that the reconstructed JPEG images using the proposed 3L-HQC MMS LDPC decoder have similar image quality to that of a PEG-based decoder and much better quality than that of the 2L-HQC-based decoder. This clearly demonstrates the enhancement made by the proposed 3L-HQC decoder with layered permutation of sub-matrices over conventional 2L-HQC decoders.

Table 7.3 Comparison of the Quality of Original and Reconstructed JPEG Images for Various Decoders

	Original Image	Reconstructed Image at E_b/N_o of 3.5 dB Using 2304-Bit LDPC Code		
		Presented Decoder	PEG-Based Decoder	2L-HQC-Based Decoder
Lena				
Golden Gate				

REFERENCES

[1] P. Ma, et al., High-rate LDPC codes in image transmission over Rayleigh fading channel, 1st IEEE Consumer Communications and Networking Conference, IEEE, Las Vegas, Nevada, 2004.

[2] P. Lingling, et al., LDPC-based iterative joint source-channel decoding for JPEG2000, IEEE Trans. Image Process. 16 (2) (2007) 577–581.

[3] N. Rahnavard, F. Fekri, New results on unequal error protection using LDPC codes, IEEE Communications Letters 10 (1) (2006) 43–45.

[4] C.Zhong, J.P. Havlicek. LDPC codes for robust transmission of images over wireless channels. in: Thirty-Fifth Asilomar Conference on Signals, Systems and Computers, California, USA, 2001.

[5] A.H.M. Almawgani, M.F.M. Salleh, Performance optimization of hybrid combination of LDPC and RS codes using image transmission system over fading channels, Eur. J. Sci. Res. 35 (1) (2009) 34–42.

[6] G. Kong, S. Choi, Performance evaluation of the reed solomon and low density parity check codes for blu-ray disk channels, Jpn. J. Appl. Phys. 49 (8) (2010) 1–3.

[7] N. Xie, T. Zhang, E.F. Haratsch, Improving burst error tolerance of LDPC-centric coding systems in read channel, IEEE Trans. Magnet. 46 (3) (2010) 933–941.

[8] S. Sankaranarayanan, A. Kuznetsov, D. Sridhara. On the concatenation of LDPC and RS codes in magnetic recording systems. in: IEEE Globecom Workshops, Washington DC, 2007.

[9] Z. Bo, et al. Non-binary LDPC codes vs. Reed-Solomon codes. in: Information Theory and Applications Workshop, San Diego, USA, 2008.

[10] G. Baruffa, P. Micanti, F. Frescura, Error protection and interleaving for wireless transmission of JPEG 2000 images and video, IEEE Trans. Image Process. 18 (2) (2009) 346–356.

[11] A. Skodras, C. Christopoulos, T. Ebrahimi, The JPEG 2000 still image compression standard, IEEE Signal Process. Mag. 18 (5) (2001) 36–58.

[12] T. ETSI, 136 212 LTE. Evolved Universal Terrestrial Radio Access (EUTRA), 2016.

[13] Q. Huynh-Thu, M. Ghanbari, Scope of validity of PSNR in image/video quality assessment, Electronics Lett. 44 (13) (2008) 800–801.

[14] W. Yi, C. Qiqiang. A novel quadratic type variational method for efficient salt-and-pepper noise removal. in: IEEE International Conference on Multimedia and Expo, Singapore, 2010.

[15] K.K.V. Toh, N.A.M. Isa, noise adaptive fuzzy switching median filter for salt-and-pepper noise reduction, IEEE Signal Process. Lett. 17 (3) (2010) 281–284.

[16] V.A. Chandrasetty, S.M. Aziz, Resource efficient LDPC decoders for multimedia communication, Integr. VLSI J. 48 (2015) 213–220.

Prospective LDPC applications

8.1 WIRELESS COMMUNICATION

LDPC codes have been adopted in various wireless communication standards because of its unique advantages. LDPC codes can be designed for varied code lengths, rates, and regularity while still achieving the performance that is expected by a wide range of wireless applications. Some of the advanced and next-generation applications that incorporate LDPC codes are as follows:

- *GEO-Mobile Radio (GMR):*

  GMR is standard for satellite telephony that is specified at ETSI (European Telecommunications Standardization Institute) [1,2]. This standard is similar to GSM (Global System for Mobile) communication and is characterized by low throughput requirement of less than 1 Mbps [3]. It uses LDPC codes for FEC and the code length for GMR varies from 950 to 11136 with over five code rates.

- *Wireless Regional Area Network (WRAN):*

  WRAN is an IEEE 802.22 standard for wireless broadband access that uses the white spaces between the occupied channels in television frequency spectrum [4]. WRAN uses cognitive radio techniques for dynamically configuring to use the best wireless channels that are available in the area. This standard is highly suitable for providing broadband access to low population density areas worldwide [5]. WRAN standard uses LDPC codes for error correction with ½, ⅔, ¾, and ⅚ as code rates and over 21 different code lengths.

- *LTE Mobile Communication:*

  Long Term Evolution (LTE) is a wireless communication standard developed for high-speed data for mobile phones and data terminals [6]. The LTE features expand over the 4G network and are being considered for 5G as LTE-Advanced. The high-performance LDPC codes potentially bolster the LTE standard to achieve the desired high-speed data rates [7]. The code lengths for LTE range from 40 to as high as 6144.

Resource Efficient LDPC Decoders. DOI: https://doi.org/10.1016/B978-0-12-811255-7.00008-3

- *Deep space communication:*

  The NASA (National Aeronautics and Space Administration) in association with CCDS (Consultative Committee for space Data Systems) has published a standard for near earth and deep space communication system [8]. This standard has an option to use LDPC codes for error correction [9]. The document recommends one set of LDPC codes for near-earth communication (code length: 8176) and other codes are optimized for deep-space communication (code length: 1024, 4096, and 16384).

- *Next Generation broadcasting to Handheld (NGH):*

  The DVB Next Generation broadcasting system to Handheld (DVB-NGH) device is a draft ETSI standard [10]. The standard defines next generation transmission systems for digital terrestrial and satellite broadcasting for handheld terminals [11]. This standard is classified into different profiles based on the number of antennas and tuners at the receiver. DVB-NGH uses LDPC codes with code length 4320, 8640, and 16200 with varied code rates.

8.2 OPTICAL COMMUNICATION

Optical communication has been a promising channel in catering for the bandwidth needs of the ever-expanding base of internet users and their demand for accessing high-definition (HD) media content over the medium [12,13]. Optical fibers use photons for transmitting information over optical cables and have clear advantages over wires and co-axial mediums in terms of bandwidth capacity, maintenance, size, and immunity to electromagnetic signals [14]. However, the transmitted optical information can have errors at the receiver due to limitations in the detection mechanism. The errors could be as a result of inter-symbol interference in the channel, shot noise or even thermal noise [15]. LDPC codes have been instrumental in augmenting the merits of optical communication by correcting the errors introduced in the optical communication. Some of the cutting-edge optical communication applications are as follows:

- *Optical Transport Network (OTN):*

  The International Telecommunication Union (ITU) defines the OTN as a set of optical network elements connected by optical fiber links that are able to provide certain functionalities such as transport, multiplexing, routing, management, supervision, and survivability of optical channels that carry information [16]. OTN offers several advantages over SONET/SDH, like stronger FEC, multiple levels of tandem connections, transparent transport of client signals, and switching scalability [17]. Currently Reed-Solomon RS (255,239) codes are implemented as FEC in OTN [18]. The coding gain provided by FEC enables the optical path to cross over more transparent optical network elements. To further improve the performance, LDPC codes can be explored as a potential FEC system in OTN elements.

- *Quantum Key Distribution (QKD):*

  QKD uses quantum properties for exchanging secret information or a cryptographic key over an insecure channel [19,20]. This key can then be used to encrypt messages that are being communicated over the same channel. The security of QKD is characterized based on the fundamentals of quantum mechanics that an act of measuring a quantum system disrupts the system. Therefore, when an eavesdropper tries to intercept the quantum channel, it will inevitably leave traces of such an act in the system. This makes the QKD system very strong against any powerful security threats on the information exchanged over the quantum channel typically an optical fiber. LDPC codes are used for augmenting the performance of QKD security over optical fiber channels [21].

- *Optical Wireless Communication (OWC):*

  OWC refers to the transmission of the visible, Infrared (IR), and ultraviolet (UV) spectral band as optical carriers over an unguided propagation medium, such as free-space [22]. OWC systems have large optical bandwidth and can achieve very high data rates to transmit data for long distances of several kilometers. Depending on the transmission range, OWC is applicable in various applications including intra/inter chip communication, inter-vehicular communication, inter-building, and inter-satellite communication [23]. The transmission medium in OWC is typically the atmosphere, which exposes the optical signals to several noise conditions, such as thermal noise, transmitter noise, shot noise, etc. High performance LDPC codes can best fit in OWC for recovering signal errors at the receiver in OWC systems.

8.3 FLASH MEMORY DEVICES

Flash memory devices store large quantities of data in a small area using stacked memory cells. The density of such cells is increasing substantially to cater for the demand of storing more data in a smaller area. In the quest to achieve this goal, the flash memory technology is being constrained with multilevel cells (MLC), where a single memory cell is designed to store more bits of data by using multiple logic levels [24]. Traditionally, a single memory cell could hold only one non-zero charge level for storing 1-bit of data. As the memory technology is moving towards MLC and more complex forms in 3D structures [25], there is an even greater challenge to counter errors introduced while storing and retrieving data from such devices. The increase in the number of levels (means reduced distance between the levels) in the cells may result in variations in cell behavior because of cell-to-cell interference. It also reduces signal-to-noise ratio while reading data from the memory stack [26]. Other factors such as random telegraph noise (RTN), retention noise due to electrons leaking from floating gate, and wear/tear of the cells substantially affect the reliability of stored data in the memory.

Therefore, using an efficient error correction technique is highly crucial for safeguarding and recovering critical data from the storage devices. LDPC codes have been largely promising in dealing the data integrity issues faced by memory devices using ultra innovative storage solutions [26]. LDPC codes are already being used as a reliable error correction technique in a number of storage devices in the industry. For example, LDPC codes are used as media error correction in SSD [27] cards SanDisk X400 and Seagate Nytro.

A flash storage device typically consists of a memory controller and flash memory stack in a single package. The controller is the common interface between the host and the flash memory. It moderates and supports various protocols (USB, PCI, UFS, etc.,) involved in the host communication with the device. A block diagram of a simple memory device package is shown the Fig. 8.1.

For writing data to the flash storage in a memory device, the host data is first buffered through a host interface. The CPU controls the LDPC encoding operation on the host data read from memory buffers. The encoded data is then buffered before storing it in the flash memory via flash interface. A similar process is followed for reading data from flash storage. The data read from the flash memory is decoded using a LDPC decoder. Errors that are introduced by faulty flash storage cells are corrected by the LDPC decoder using multilevel error correction schema [28]. The corrected data is then sent to the host via the memory buffers and host interface. For soft-decision decoding, the soft information is obtained by performing multiple reads with varied word-line voltages. The hard-decision decoding scheme is first used for error correction. In case of correction failure, soft-decision decoding [28] with higher levels of sampling and quantization scheme is attempted in stages for correcting the errors.

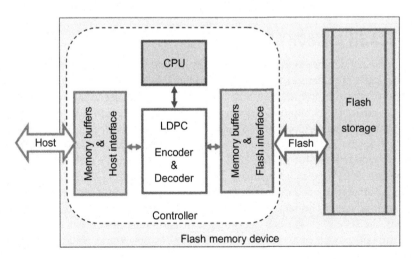

FIGURE 8.1

Internal block diagram of a flash memory device.

REFERENCES

[1] T. ETSI. 101 376-1-3 V3. 1.1 (2009-07) Technical Specification GEO-Mobile Radio Interface Specifications (Release 3) Third Generation Satellite Packet Radio Service.

[2] P. Chini, G. Giambene, S. Kota, A survey on mobile satellite systems, Int. J. Satellite Commun. Netw. 28 (1) (2010) 29−57.

[3] S.D. Ilčev, Mobile Satellite Antenna Systems, in Global Mobile Satellite Communications Theory, Springer, 2017, pp. 301−392.

[4] IEEE Standard for Information Technology−Telecommunications and information exchange between systems - Wireless Regional Area Networks (WRAN)−Specific requirements - Part 22: Cognitive Wireless RAN Medium Access Control (MAC) and Physical Layer (PHY) Specifications:Policies and Procedures for Operation in the TV Bands - Amendment 2: Enhancement for Broadband Services and Monitoring Applications. IEEE Std 802.22b-2015 (Amendment to IEEE Std 802.22-2011 as amended by IEEE Std 802.22a-2014), 2015: p. 1−299.

[5] J. Kanti, G.S. Tomar, A. Bagwari, An improved-two stage detection technique for IEEE 802.22 WRAN, Optik-Int. J. Light Electron Optics 140 (2017) 695−708.

[6] T. ETSI. 136 212 LTE. Evolved Universal Terrestrial Radio Access (EUTRA), 2016.

[7] A.L. Ortega-Ortega, J.F. Bravo-Torres. Combining LDPC codes, M-QAM modulations, and IFDMA multiple-access to achieve 5G requirements. in Electronics, Communications and Computers (CONIELECOMP), 2017 International Conference on. 2017: IEEE.

[8] M. Kearney, K. Tuttle, J. Afarin, NASA Report to the CCSDS Management Council, 2014.

[9] A. Babuscia, D. Divsalar, K.-M. Cheung. CDMA communication system for mars areostationary relay satellite. in Aerospace Conference, 2017 IEEE. 2017: IEEE.

[10] BlueBook, Next Generation broadcasting system to Handheld, physical layer specification (DVB-NGH). ETSI, 2012.

[11] D. Gómez-Barquero, et al., DVB-NGH: The next generation of digital broadcast services to handheld devices, IEEE Trans. Broadcast. 60 (2) (2014) 246−257.

[12] H. Venghaus, N. Grote, Fibre Optic Communication: Key Devices, Vol. 161, Springer, 2017.

[13] H. Meyr, M. Moeneclaey, S.A. Fechtel, Digital Communication Receivers: Synchronization, Channel Estimation, and Signal Processing, Wiley Online Library, 1998.

[14] P. Gallion, Applications to Optical Communication, Micro-and Nanophotonic Technologies, 2017, pp. 291−332.

[15] G. Li, X. Zhou. Next-Generation Optical Communication: Components, Sub-Systems, and Systems VI. in Proc. of SPIE Vol. 2017.

[16] I.T.S. Sector, Optical transport network (OTN), ITU-T Rec. G 709 (2003) Y1331.

[17] M.C. Kapse, S. Shriramwar, Forward Error Correction (FEC) computation in Optical Transmission Systems, 2017.

[18] V. Houtsma, D. van Veen, E. Harstead, Recent progress on standardization of next-generation 25, 50, and 100G EPON, J. Lightwave Technol. 35 (6) (2017) 1228−1234.

[19] J. Ouellette, Quantum key distribution, Industrial Phys. 10 (6) (2004) 22−25.

[20] S.-K. Liao, et al., Long-distance free-space quantum key distribution in daylight towards inter-satellite communication, Nat. Photonics 11 (8) (2017) 509−513.

[21] S.J. Johnson, V.A. Chandrasetty, A.M. Lance. Repeat-accumulate codes for reconciliation in continuous variable quantum key distribution. in Communications Theory Workshop (AusCTW), 2016 Australian. 2016: IEEE.

[22] S. Arnon, Optimization of urban optical wireless communication systems, IEEE Trans. Wireless Commun. 2 (4) (2003) 626–629.

[23] K.A. Mekonnen, et al. Over 40 Gb/s dynamic bidirectional all-optical indoor wireless communication using photonic integrated circuits. in: Optical Fiber Communications Conference and Exhibition (OFC), 2017, 2017: IEEE.

[24] E.-S. Choi, S.-K. Park. Device considerations for high density and highly reliable 3D NAND flash cell in near future. in: Electron Devices Meeting (IEDM), 2012 IEEE International. 2012: IEEE.

[25] C.-Y. Lu, Future prospects of NAND flash memory technology—the evolution from floating gate to charge trapping to 3D stacking, J. Nanosci. Nanotechnol. 12 (10) (2012) 7604–7618.

[26] J. Guo, et al., Flexlevel NAND flash storage system design to reduce LDPC latency, IEEE Trans. Comput. Aided Design Integr. Circuits Syst. 36 (7) (2017) 1167–1180.

[27] L. Zuolo, et al., Design Trade-Offs for NAND Flash-Based SSDs, in Solid-State-Drives (SSDs) Modeling, Springer, 2017, pp. 67–97.

[28] L. Zuolo, et al., LDPC soft decoding with improved performance in 1X-2X MLC and TLC NAND flash-based solid state drives, IEEE Trans. Emerging Top. Comput. (2017).

Sample C-Programs and MATLAB models for LDPC code construction and simulation

```
%%%%%%%%%%%%%%%%%%%%%%%%%%%%%%%%%%%%%%%%%%%%%%%
% DESIGN     :    GenHqcLDPC
% DESIGNER   :    Vikram A. Chandrasetty
% EMAIL      :    vikramac@ieee.org
% PROGRAM    :    MATLAB
%
% DESCRIPTION:
% Top level MATLAB model to generate Hierarchical
% Quasi-Cyclic LDPC matrix.
%
% Copyright (c) 2017, www.vikramac.com
%%%%%%%%%%%%%%%%%%%%%%%%%%%%%%%%%%%%%%%%%%%%%%%

function [HQC_LDPC] = GenHqcLDPC(RandMat, ParallelNodes, BaseCL,
ExpFactor)

load(RandMat);

BaseSize      = ParallelNodes; %18;
CoreSize      = BaseCL; %648;
CodeLength    = CoreSize*ExpFactor;

fprintf('Generating LDPC Matrix of Code Length: %d...\n', CodeLength);

%%Formation of Core Matrix
%Expand the Permuted matrix for the given Expansion Factor
ExT1 = Gen_Ex_Rand(T1, ExpFactor);
ExT2 = Gen_Ex_Rand(T2, ExpFactor);
ExT3 = Gen_Ex_Rand(T3, ExpFactor);

PermShift = 1;

%Layer 0
Core1_0 = ShiftR(ExT1,        -0*PermShift);
Core1_1 = ShiftR(ExT1,        -1*PermShift);
Core1_2 = ShiftR(ExT1,        -2*PermShift);
```

```matlab
Core1_3 = ShiftR(ExT1,        -3*PermShift);
Core1_4 = ShiftR(ExT1,        -4*PermShift);
Core1_5 = ShiftR(ExT1,        -5*PermShift);

%Layer 1
Core2_0 = ShiftR(ExT2,        -0*PermShift);
Core2_1 = ShiftR(ExT2,        -1*PermShift);
Core2_2 = ShiftR(ExT2,        -2*PermShift);
Core2_3 = ShiftR(ExT2,        -3*PermShift);
Core2_4 = ShiftR(ExT2,        -4*PermShift);
Core2_5 = ShiftR(ExT2,        -5*PermShift);

%Layer 2
Core3_0 = ShiftR(ExT3,        -0*PermShift);
Core3_1 = ShiftR(ExT3,        -1*PermShift);
Core3_2 = ShiftR(ExT3,        -2*PermShift);
Core3_3 = ShiftR(ExT3,        -3*PermShift);
Core3_4 = ShiftR(ExT3,        -4*PermShift);
Core3_5 = ShiftR(ExT3,        -5*PermShift);

%Matrix unexpaneded with shifted Identity matrix
UnExpMat = [Core1_0, Core1_1, Core1_2, Core1_3, Core1_4, Core1_5;
            Core2_0, Core2_1, Core2_2, Core2_3, Core2_4, Core2_5;
            Core3_0, Core3_1, Core3_2, Core3_3, Core3_4, Core3_5];

%3 Level HQC matrix
HQC_LDPC = Gen_RI_Shift(UnExpMat, BaseSize);

%Verify regularity (3,6) and Girth (>4)
FindRegularity(HQC_LDPC);
FindGirth4(HQC_LDPC);

function [M_Ex_Rand] = Gen_Ex_Rand(Rand, ExFactor)
Len = length(Rand);
M_Ex_Rand = zeros(Len*ExFactor);
Ex_Rand_Len = length(M_Ex_Rand);

Spacing = 1;
SpaceSpan = zeros(Ex_Rand_Len);
SpaceSpan(1,1) = Spacing;

for L = 2:Ex_Rand_Len
      Spacing = Spacing + ExFactor;
      SpaceSpan(1,L) = Spacing;
end

for c = 1:Len
      pos = find(Rand(:,c)>0);
      M_Ex_Rand(pos,SpaceSpan(1,c)) = Rand(pos, c);
end
```

```
for e = 2:ExFactor
        for r = 1:Len
                for c = 1:((Len*ExFactor)-ExFactor+1)
                        M_Ex_Rand((Len*(e-1))+r, c+e-1) = M_Ex_Rand(r, c);
                end
        end
end

function [M_RI_Shift] = Gen_RI_Shift(Matrix, BaseSize)
[R, C] = size(Matrix);

M_RI_Shift = zeros(R*BaseSize, C*BaseSize);
Imat = eye(BaseSize);
IZero = zeros(BaseSize);

CnodeStart = 1;
CnodeEnd = BaseSize;
VnodeStart = 1;
VnodeEnd = BaseSize;

for Cnode = 1:R

        for Vnode = 1:C

                if(Matrix(Cnode, Vnode) == 0)
                   M_RI_Shift(CnodeStart:CnodeEnd, VnodeStart:
VnodeEnd) = IZero;
                else
                   M_RI_Shift(CnodeStart:CnodeEnd, VnodeStart:
VnodeEnd) = ShiftR(Imat, Matrix(Cnode, Vnode));
                end

                VnodeStart = VnodeEnd + 1;
                VnodeEnd     = VnodeEnd + BaseSize;

        end

        CnodeStart = (Cnode*BaseSize)+1;
        CnodeEnd     = (Cnode+1)*BaseSize;
        VnodeStart = 1;
        VnodeEnd     = BaseSize;

end

function [ShiftMat] = ShiftR(Mat, sf)
ShiftMat = circshift(Mat, -sf);

function FindRegularity(H)

Vnode = sum(mod(sum(H),  3));
Cnode = sum(mod(sum(H'), 6));
```

```
if(Vnode == 0 && Cnode == 0)
        fprintf('Regular (3, 6)\n');

else
        fprintf('FAILED: Irregular\n');

end

function FindGirth4(H)
[M,N] = size(H);

O = H*H';
for i = 1:M
        O(i,i) = 0;
end

girth = zeros(1,M);
for i = 1:M
        girth(i) = max(O(i,:));

end

girth4 = max(girth);
if girth4 < 2
    fprintf('No girth 4\n')
else
    fprintf('FAILED: The H matrix has girth 4\n') % Provide the test
result.

    end
```

```
/******************************************************************************

*       DESIGN    :        LdpcHdlAutoFPA
*       DESIGNER :        Vikram A. Chandrasetty
*       EMAIL     :        vikramac@ieee.org
*       PROGRAM   :        C program (MEX compiler in MATLAB)
*
*       DESCRIPTION :
*       This application is used to automatically generate
*       Verilog HDL code for a given LDPC matrix. The HDL
*       code consists of the network interconnects of the
*       LDPC matrix
*
*       Copyright (C) 2017, www.vikramac.com
******************************************************************************/
```

```c
#include <math.h>
#include <mex.h>
#include <matrix.h>
#include <stdlib.h>
#include <string.h>
#include <stdio.h>

//Input Macros
#define ALGNAME      prhs[0]
#define HMAT         prhs[1]
#define LLRQUANT     prhs[2]
#define EXTMSGLEN    prhs[3]

//Other Macros
#define H            (int)Hmat

//HDL Strings
#define NUMINPORT    5
#define NUMOUTPORT   2
#define NUMCNPORT    13
#define NUMVNPORT    12
#define NUMPARAMS    4
#define STRLENGTH    50

#define MOD_NAME_KEY     "LDPC_Net_H"
#define CN_NAME_KEY      "Cnode_"
#define VN_NAME_KEY      "Vnode_"
#define PARITYWIRE       "wParityOut"
#define FILE_EXT         ".v"
#define TIME_SCALE       "timescale 1ns / 1ps"

//Global Variables
char*       AlgName;
double*     Hmat;
int         LlrQ;
int         ExtMsgLen;
int         numCheckNodes;
int         numVarNodes;
int         CnodeDegree;
int         VnodeDegree;
int         *HmatWires;

static FILE    *fp = NULL;

char    InputPorts[NUMINPORT][STRLENGTH] = {"Clock", "Reset", "Load",
"Start", "LLRin"};
char    OutputPorts[NUMOUTPORT][STRLENGTH] = {"DecodedOut",
"ParityOut"};
char    CnodePortList[NUMCNPORT][STRLENGTH] = {"In0", "In1", "In2",
"In3", "In4","In5", "Out0", "Out1", "Out2", "Out3", "Out4", "Out5",
"ParityOut"};
```

```c
char      VnodePortList[NUMVNPORT][STRLENGTH] = {"Clk", "Rst", "Load",
"Start", "LLR", "Vin0", "Vin1", "Vin2", "Vout0", "Vout1", "Vout2",
"BitOut"};
char      ParamNames[NUMPARAMS][STRLENGTH]        =        {"CODE_LENGTH",
"LLR_INPUT_WIDTH", "PARITY_OUT_WIDTH", "EXT_MSG_WIDTH"};
char      *ModuleName;
char      *FileName;
char      *CnName;
char      *VnName;

static void PreProcessInfo(){
    int c, v, e, addr;
    //Allocate memory for storing wire names
    HmatWires = calloc( numVarNodes*numCheckNodes, sizeof(int));

    e = 0;
    //Initialize    memory    with    wire    names:    {EdgeNumber,
    CheckNodeNumber}
    for(c = 0; c < numCheckNodes; c++){
        //e = 0;
        for(v = 0; v < numVarNodes; v++){
            addr = v*numCheckNodes + c;
            if(H[addr] > 0){
                HmatWires[addr] = e++;
            }
        }
    }

    //Create Module Name
    ModuleName = calloc(STRLENGTH, sizeof(char));
    sprintf(ModuleName,   "%s%d_%s",   MOD_NAME_KEY,   numVarNodes,
    AlgName);

    //Create File Name
    FileName = calloc(STRLENGTH, sizeof(char));
    sprintf(FileName, "%s%s", ModuleName, FILE_EXT);

    //Create Check Node Name
    CnName = calloc(STRLENGTH, sizeof(char));
    sprintf(CnName, "%s%s", CN_NAME_KEY, AlgName);

    //Create Varaiable Node Name
    VnName = calloc(STRLENGTH, sizeof(char));
    sprintf(VnName, "%s%s", VN_NAME_KEY, AlgName);

    //PrintMatrix(HmatWires, "Wires");
    return;

}
```

```
static void PrintHeader(){
      fprintf(fp, "/////////////////////////////////////////////\n");
      fprintf(fp, "// ********** Autogenerated File - DO NOT MODIFY
      **********\n");
      fprintf(fp, "// Module Name: %s\n", ModuleName);
      fprintf(fp, "// \n");
      fprintf(fp, "// Description: LDPC graph inter-connection module
      for \n");
      fprintf(fp, "// Fully-Parallel decoder architecture \n");
      fprintf(fp, "// \n");
      fprintf(fp, "/////////////////////////////////////////////\n\n");
      return;

}
static void PrintTimescale() {
      fprintf(fp, "%s\n", TIME_SCALE);
}
static void GenModHeader(){
      int i;

      fprintf(fp, "module %s (\n", ModuleName);

      for(i = 0; i < NUMINPORT; i++){
            fprintf(fp, "\t%s,\n", InputPorts[i]);
      }

      for(i = 0; i < NUMOUTPORT; i++){
            fprintf(fp, "\t%s", OutputPorts[i]);
            if(i + 1 < NUMOUTPORT)
                  fprintf(fp, ",\n");
      }

      fprintf(fp, "\n\t);\n\n");
      return;

}

static void GenParameter(){
      int i = 0;

      //Code Length
      fprintf(fp,  "parameter  %s  =  (%d)-1;\n",  ParamNames[i],
      numVarNodes);

      //LLR input width
      fprintf(fp,  "parameter  %s  =  (%d)-1;\n",  ParamNames[++i],
      numVarNodes*LlrQ);
```

```
        //Parity output width
        fprintf(fp, "parameter %s = (%d)-1;\n", ParamNames[++i],
        numCheckNodes);

        //Extrinsic message length
        fprintf(fp, "parameter %s = (%d)-1;\n", ParamNames[++i],
        ExtMsgLen);

        fprintf(fp, "\n");
        return;

}

static void GenIOPorts(){
        int i;

        i = 0;
        //Clock
        fprintf(fp, "\ninput %s;", InputPorts[i]);
        //Reset
        fprintf(fp, "\ninput %s;", InputPorts[++i]);
        //Load
        fprintf(fp, "\ninput [%s:0] %s;", ParamNames[0], InputPorts
        [++i]);
        //Start
        fprintf(fp, "\ninput %s;", InputPorts[++i]);
        //LLR input
        fprintf(fp, "\ninput [%s:0] %s;", ParamNames[1], InputPorts
        [++i]);

        i = 0;
        //DecodedOut
        fprintf(fp, "\n\noutput [%s:0] %s;", ParamNames[0], OutputPorts
        [i]);
        //ParityOut
        fprintf(fp, "\noutput %s;", OutputPorts[++i]);

        fprintf(fp, "\n\n");
        return;

}

static void GenWires(){
        int c, v, addr;

        //Wires for ParityOut
        fprintf(fp, "\nwire [%s:0] %s;", ParamNames[2], PARITYWIRE);
```

```
        for(c = 0; c < numCheckNodes; c++){
            for(v = 0; v < numVarNodes; v++){
                addr = v*numCheckNodes + c;
                if(H[addr] > 0){
                    if(ExtMsgLen > 1){
                        fprintf(fp,"\nwire      [%s:0]      A%d;",
                        ParamNames[3], HmatWires[addr]);
                        fprintf(fp,"\nwire      [%s:0]      B%d;",
                        ParamNames[3], HmatWires[addr]);
                    }
                    else {
                        fprintf(fp,    "\nwire   A%d;",    HmatWires
                        [addr]);
                        fprintf(fp,    "\nwire   B%d;",    HmatWires
                        [addr]);
                    }
                }
            }
        }
        fprintf(fp, "\n\n");
        return;

}

static void GenCheckNodeInst(){
        int c, v, addr, netNum;

        for(c = 0; c < numCheckNodes; c++){
            fprintf(fp, "\n%s CN%d\t(", CnName, c);

            netNum = 0;
            //Print input ports
            for(v = 0; v < numVarNodes; v++){
                addr = v*numCheckNodes + c;
                if(H[addr] > 0){
                    fprintf(fp,".%s({B%d}),   ",   CnodePortList
                    [netNum++], HmatWires[addr]);
                }
            }

            //Print output ports
            for(v = 0; v < numVarNodes; v++){
                addr = v*numCheckNodes + c;
                if(H[addr] > 0){
                    fprintf(fp,".%s({A%d}),   ",   CnodePortList
                    [netNum++], HmatWires[addr]);
                }
            }
```

```
                    //Parity Out Port
                    fprintf(fp,   ".%s(%s[%d])",   CnodePortList[netNum++],
                    PARITYWIRE,c);

                    //Complete the instance syntax
                    fprintf(fp, ");");
          }

          fprintf(fp, "\n\n");
          return;


}

static void GenVarNodeInst(){
          int c, v, addr, i, netNum;

          for(v = 0; v < numVarNodes; v++){
                    fprintf(fp, "\n%s VN%d\t(", VnName, v);

                    i = 0;
                    netNum = 0;
                    //Print control ports
                    fprintf(fp,".%s(%s),",VnodePortList[netNum++],
                    InputPorts[i]);        //Clock
                    fprintf(fp,".%s(%s),",VnodePortList[netNum++],
                    InputPorts[++i]); //Reset
                    fprintf(fp,".%s(%s[%d]),",VnodePortList[netNum++],
                    InputPorts[++i], v);//Load
                    fprintf(fp,".%s(%s),",VnodePortList[netNum++],
                    InputPorts[++i]);        //Start
                    fprintf(fp,".%s(%s[%d:%d]),",VnodePortList[netNum++],
                    InputPorts[++i],
                    ((v + 1)*LlrQ)-1, v*LlrQ);   //LLR   Input

                    //Print input ports
                    for(c = 0; c < numCheckNodes; c++){
                              addr = v*numCheckNodes + c;
                              if(H[addr] > 0){
                                        fprintf(fp,".%s({A%d}), ", VnodePortList
                                        [netNum++], HmatWires[addr]);
                              }
                    }

                    //Print output ports
                    for(c = 0; c < numCheckNodes; c++){
                              addr = v*numCheckNodes + c;
                              if(H[addr] > 0){
```

```
                                fprintf(fp,".%s({B%d}),   ",  VnodePortList
                                [netNum++], HmatWires[addr]);
                    }
            }

            //Print Decoded BitOutput
            fprintf(fp,    ".%s(%s[%d])",   VnodePortList[netNum++],
            OutputPorts[0], v);

            //Complete the instance syntax
            fprintf(fp, ");");
        }
        fprintf(fp, "\n\n");
        return;

}

static void GenParityLogic(){
        int c;

        fprintf(fp, "assign %s = \n\t", OutputPorts[1]); //ParityOut

        for(c=0; c<numCheckNodes; c++){
                if(c == 0)
                        fprintf(fp, " %s[%d]", PARITYWIRE, c);
                else
                        fprintf(fp, "\n\t| %s[%d]", PARITYWIRE, c);
        }

        fprintf(fp, ";\n\n");
        fprintf(fp, "endmodule\n");
        return;

}

static void GenInstances(){
    GenCheckNodeInst();
    GenVarNodeInst();
    GenParityLogic();
    return;

}

static void GenDummyCnodeMod(){
    printf("Generating Check Node Module....\n");
    return;

}
```

```c
static void GenDummyVnodeMod(){
      printf("Generating Variable Node Module....\n");
      return;

}
static void GenDummyNodeModules(){
      GenDummyCnodeMod();
      GenDummyVnodeMod();
      return;

}
static void GenTestBenches(){
      printf("Generating Testbenches....\n");
      return;

}
static void GenCnodeTb(){
      return;

}
static void GenVnodeTb(){
      return;

}
static void* OpenFile(char* fileName){
      FILE* pFile = NULL;
      printf("Opening file to write HDL : %s\n", fileName);
      pFile = fopen(fileName, "w");
      if(pFile == NULL)
            mexErrMsgTxt("Could not open file\n");

      return pFile;

}
static void CloseFile(void* pFile){
      fclose(pFile);
      return;

}
static void CleanUp(){
      free(ModuleName);
      free(FileName);
      free(VnName);
      free(CnName);
      return;

}
```

```
//Debug API
static void PrintMatrix(int* Matrix, char* Name){
      int c, v;
      printf("Printing Matrix: %s\n", Name);
      for (c = 0; c < numCheckNodes; c++){
            for (v = 0; v < numVarNodes; v++){
                  printf("%4d ", Matrix[v*numCheckNodes + c]);
            }
            printf("\n");
      }
}


static void GenerateHDL(){
    //printf("Generating HDL ...\n");

    //Open HDL file to write
    fp = (FILE*)OpenFile(FileName);

    PrintHeader();
    PrintTimescale();

    GenModHeader();
    GenParameter();
    GenIOPorts();
    GenWires();
    GenInstances();

    GenDummyNodeModules();
    GenTestBenches();

    //Close the HDL file
    CloseFile(fp);
    fp = NULL;
    return;

}

///////////// Call MEX function /////////////
void mexFunction(
            int            nlhs,
            mxArray        *plhs[],
            int            nrhs,
            const mxArray  *prhs[] ) {
```

```
//Local Variables
int c, v;

if(nrhs != 4) {
      printf("Usage:LdpcHdlAutoFPA(AlgorithmName,      Hmatrix,
      LlrQuant, ExtQuant)\n");
      return;
}


//Load Input data
AlgName          = mxArrayToString(ALGNAME);
Hmat             = mxGetPr(HMAT);
LlrQ             = *mxGetPr(LLRQUANT);
ExtMsgLen        = *mxGetPr(EXTMSGLEN);

//Extract Information
numCheckNodes = mxGetM(HMAT);
numVarNodes   = mxGetN(HMAT);

VnodeDegree = 0;
for(c = 0; c < numCheckNodes; c++){
      if(H[c] > 0)
               ++VnodeDegree;
}

CnodeDegree = 0;
for(v = 0; v < numVarNodes; v++){
   if(H[v*numCheckNodes] > 0)
      ++CnodeDegree;
}

printf("LDPC matrix (H): \n\t Code Length: %d \n\t Code Rate: %
2.1f \n\t Check Node Degree: %d \n\t Variable Node Degree: %d\n",
numVarNodes, ((float)numCheckNodes/numVarNodes), CnodeDegree,
VnodeDegree);

PreProcessInfo();

GenerateHDL();

CleanUp();
free(HmatWires);
return;

}
```

```
%%%%%%%%%%%%%%%%%%%%%%%%%%%%%%%%%%%%%%%%%%%%%%%%
% DESIGN     :   SimLDPC
% DESIGNER   :   Vikram A. Chandrasetty
% EMAIL      :   vikramac@ieee.org
% PROGRAM    :   MATLAB
%
% DESCRIPTION:
% Top level MATLAB model to simulate different LDPC
% decoding algorithms and also gererate vectors for
% FPGA emulation.
%
% Copyright (c) 2017, www.vikramac.com
%%%%%%%%%%%%%%%%%%%%%%%%%%%%%%%%%%%%%%%%%%%%%%%

function SimLDPC(H_Mat, Algorithm, MaxIter, SnrLevels, MaxBitErrors,
Resolution, sim_fpga)

load(H_Mat);      % Load H and EncodedData
Rate = 0.5;       % ASUMING Rate of the LDPC code
SimmType = 0;
CodeLen = length(H(1,:));

%Initialize parameters
ResBER = 1/(10^Resolution);

fprintf('Simulation: MaxDataErr(%d), Resolution(% 2.0e), CodeLength
(%d)\n',... MaxBitErrors, ResBER, CodeLen);
%Simulation Stop flags
SimmStop = 0;

%FPGA testing parameters
ComPort = 'COM1';
BaudRate = 115200;

quantBits = 4; %Default 4-bit quantization
if(str2num(Algorithm(4)) < 10 )
  quantBits = str2num(Algorithm(4));
end

%%% Choose the Algorithm
switch lower(Algorithm)
      case 'ucbm' %Uncoded BPSK
            AlgoNum = 0;
      case {'spa4'} %Sum Product Algorithm
            AlgoNum = 1;
      case {'bfo4'} %Bit Flip Algorithm
            AlgoNum = 2;
```

```
      case   {'msof','mso2','mso3','mso4','mso5',  'mso6'}  %Min-Sum
   Algorithm
           AlgoNum = 3;
      case {'mms4'} %Modified Min-Sum 2-bit
           AlgoNum = 4;
      case    {'smp2','smp3','smp4','smp5','smp6','smp7','smp8'}    %
   Modified Message Passing
           AlgoNum = 5;
      case {'oms2',  'oms3',  'oms4'} %Optimized Min-Sum 2 bit from
   literature
           AlgoNum = 6;
   otherwise
           fprintf('Please select the LDPC Decoding Algorithm\n');
           return

end

Tag = H_Mat(1,1);
if((strcmp('H',Tag) == 0)) %No EncodedData - 'R', 'Q', 'T', 'UCBM'
     NoOfTc = 1;
     %CodeLen = length(H(1,:));
     EncodedData = zeros(NoOfTc, CodeLen);
     if(AlgoNum == 0)
       fprintf('Compensation disabled\n');
       SnrComp = 0;
     else
           SnrComp = 0.25;
     end
else %PEG Matrix with EncodedData
     fprintf('Compensation disabled\n');
     [NoOfTc, CodeLen] = size(EncodedData);
     SnrComp = 0;

end

%%% Initialize Variables
VectWidth = length(SnrLevels);
BER = zeros(1,VectWidth);
FER = zeros(1,VectWidth);
IterStat = zeros(VectWidth, MaxIter);
IterAvg = zeros(1,VectWidth);
NumFrames = zeros(1,VectWidth);

F_BER = zeros(1,VectWidth);
F_FER = zeros(1,VectWidth);
F_IterStat = zeros(VectWidth, MaxIter);
F_IterAvg = zeros(1,VectWidth);
```

```matlab
F_NumFrames = zeros(1,VectWidth);

% Simulate for each SNR Levels (dB)
for y = 1: length(SnrLevels)

    if(SimmStop)
        continue;
    end

    if(strcmpi(sim_fpga, 'simm'))
        fprintf('\nSIMM "%s" (Q:%d) SNR: %3.2fdB => ',...
            Algorithm, quantBits, SnrLevels(y));
        SimmType = 0;
    end

    if(strcmpi(sim_fpga, 'fpga'))
        fprintf('\nFPGA "%s" (Q:%d) SNR: %3.2fdB => ',...
            Algorithm, quantBits, SnrLevels(y));
        SimmType = 1;
    end

    if(strcmpi(sim_fpga, 'simm_fpga'))
        fprintf('\nSIMM_FPGA "%s" (Q:%d) SNR: %3.2fdB => ',...
            Algorithm, quantBits, SnrLevels(y));
        SimmType = 2;
    end

    %Start Timer
    startTime = clock;

    % Call C API
    [Output, iter, T_IterAvg, T_IterStat, T_BER, T_FER, T_NumFrames]
    = ... LdpcDec(H, AlgoNum, quantBits, MaxIter, (SnrLevels(y) +
    SnrComp),
    EncodedData,...
            MaxBitErrors, ComPort, BaudRate, SimmType);

    % Stop Timer
    eMin = floor(etime(clock,startTime)/60); % Elapsed Minutes
    eSec = mod(ceil(etime(clock,startTime)),60); % Elapsed Seconds

    if(strcmpi(sim_fpga, 'simm'))
        BER(y) = T_BER(1);
        FER(y) = T_FER(1);
        IterAvg(y) = T_IterAvg(1);
        NumFrames(y) = T_NumFrames(1);
        fprintf('BER: %2.1e, FER: %2.1e (% 7d), AvgIter: %3.1f
        (Time: %dm %ds)',...
```

```
                              BER(y), FER(y), NumFrames(y), IterAvg(y), eMin,
                              eSec);

                    % If resolution is sufficient
                    if(BER(y) <= ResBER)
                            SimmStop = 1;
                    end
            end

            if(strcmpi(sim_fpga, 'fpga'))
                    F_BER(y) = T_BER(1);
                    F_FER(y) = T_FER(1);
                    F_IterAvg(y) = T_IterAvg(1);
                    F_NumFrames(y) = T_NumFrames(1);
                    fprintf('BER: %2.1e, FER: %2.1e (% 7d), AvgIter: %3.1f
                    (Time: %dm %ds)',...
                            F_BER(y), F_FER(y), F_NumFrames(y), F_IterAvg(y),
                            eMin, eSec);
                    % If resolution is sufficient
                    if(F_BER(y) <= ResBER)
                            SimmStop = 1;
                    end
            end

            if(strcmpi(sim_fpga, 'simm_fpga'))
                    BER(y) = T_BER(1);
                    FER(y) = T_FER(1);
                    IterAvg(y) = T_IterAvg(1);
                    NumFrames(y) = T_NumFrames(1);

                    F_BER(y) = T_BER(2);
                    F_FER(y) = T_FER(2);
                    F_IterAvg(y) = T_IterAvg(2);
                    F_NumFrames(y) = T_NumFrames(2);

                    fprintf('BER: (%2.1e, %2.1e), FER: (%2.1e, %2.1e) [% 7d, %
                    7d], AvgIter: (%3.1f, %3.1f) (Time: %dm %ds)',...
                            BER(y), F_BER(y), FER(y), F_FER(y), NumFrames(y),
                            F_NumFrames(y), IterAvg(y), F_IterAvg(y), eMin,
                            eSec);

                    % If resolution is sufficient
                    if((BER(y) <= ResBER) && (F_BER(y) <= ResBER))
                            SimmStop = 1;
                    end
            end
```

```
% Save Data
if(strcmpi(sim_fpga, 'simm') || strcmpi(sim_fpga, 'simm_fpga'))
      simDataFile =
 strcat('SIMM_',Algorithm,'_',Tag,int2str(CodeLen),'_I',int2str
      (MaxIter));
      save (simDataFile, 'SnrLevels', 'BER', 'FER', 'IterAvg',
      'IterStat');

end

F_FREQ = 0;
if(strcmpi(sim_fpga, 'fpga') || strcmpi(sim_fpga, 'simm_fpga'))
      FpgaDataFile =
 strcat('FPGA_',Algorithm,'_',Tag,int2str(CodeLen),'_I',int2str
      (MaxIter));
      save (FpgaDataFile, 'SnrLevels', 'F_BER', 'F_FER', 'F_IterAvg',
 'F_IterStat', 'F_FREQ');
end

end

if(strcmpi(sim_fpga, 'simm') || strcmpi(sim_fpga, 'simm_fpga'))
      fprintf('\nSaving simulation data => %s.mat\n', simDataFile);
end

if(strcmpi(sim_fpga, 'fpga') || strcmpi(sim_fpga, 'simm_fpga'))
      fprintf('\nSaving FPGA test data => %s.mat\n', FpgaDataFile);
end
```

Sample Verilog HDL codes for implementation of fully-parallel LDPC decoder architecture

```
///////////////////////////////////////////////////////////////////
// Module Name      :  DUT_LDPC_FPA
// Designer         :  Vikram A. Chandrasetty
// Email            :  vikramac@ieee.org
// Program          :  Verilog HDL
//
// Description      :
// Verilog HDL top module for LDPC fully-parallel Decoder.
// 1/2 rate (3,6) 1152-bit LDPC code for LTE 4G/5G applications.
// RS232 transceiver is integrated in the design for serial
// communcation interface with the decoder.
//
// Copyright (C) 2017, www.vikramac.com
///////////////////////////////////////////////////////////////////

`timescale 1ns / 1ps
module DUT_LDPC_FPA(
    Clock,
    Reset,
    SerialIn,
    SerialOut,
    DecoderStatus
  );

//RS232 Interface Parameters
parameter CLOCK_FREQ = 100000000; //100MHz for FPGA using RS232 interface
parameter BAUD_RATE = 115200;   //Baud rate of the RS232 transceiver
parameter RS232_DATA_WIDTH = 8;   //RS232 input and output data width

//LDPC Decoder Parameters
parameter CODE_LENGTH      = 1152; //Codelength of the LDPC decoder
parameter LLR_WIDTH        = 4;   //LLR quantization
parameter MAX_ITER_WIDTH   = 4;   //Quantization of max iter (4 => max
16 counts)
```

```verilog
//IMP: Donot edit these parameters
parameter RS232_DATA_WIDTH_1    = (RS232_DATA_WIDTH)-1;
parameter CODE_LENGTH_1         = (CODE_LENGTH)-1;
parameter LLR_WIDTH_1           = (LLR_WIDTH)-1;
parameter MAX_ITER_WIDTH_1      = (MAX_ITER_WIDTH)-1;

//Ports
input Clock;  //Global clock
input Reset;   //Global reset
input SerialIn;  //Serial data input
output SerialOut;    //Serial data output
output DecoderStatus;  //Decoder activity status

//Wires
wire [RS232_DATA_WIDTH:0] SerialRegIn;
wire [RS232_DATA_WIDTH:0] SerialRegOut;
wire SerialOutAck;
wire SerialInReady;
wire SerialOutReady;

//LDPC Decoder Instance
LDPC_Serial_Comm
#(CODE_LENGTH_1, LLR_WIDTH_1, MAX_ITER_WIDTH_1, RS232_DATA_WIDTH_1)
  I_LDPC_Decoder(
    .Clock(Clock),
    .Reset(Reset),
    .SerialDataIn(SerialRegIn),
    .SerialDataInReady(SerialInReady),
    .SerialDataOutAck(SerialOutAck),
    .SerialDataOut(SerialRegOut),
    .SerialDataOutReady(SerialOutReady),
    .DecoderStatus(DecoderStatus)
    );

//RS232 Transceiver Instance
RS232_TranRec
#(RS232_DATA_WIDTH_1, CLOCK_FREQ, BAUD_RATE)
  I_RS232_TranRec(
    .Clock(Clock),
    .Reset(Reset),
    .RxIn(SerialIn),
    .RxData(SerialRegIn),
    .RxDone(SerialInReady),
    .TxOut(SerialOut),
    .TxData(SerialRegOut),
    .TxDone(SerialOutAck),
    .TxReady(SerialOutReady)
    );
endmodule
```

```
/////////////////////////////////////////////////////////////////
// Module Name    :  Vnode_MMS2
// Designer       :  Vikram A. Chandrasetty
// Email          :  vikramac@ieee.org
// Program        :  Verilog HDL
//
// Description    :
// Verilog HDL module for LUT based variable node (MMS2 algorithm)
//
// Copyright (C) 2017, www.vikramac.com
/////////////////////////////////////////////////////////////////
`timescale 1ns / 1ps
module Vnode_MMS2(
  Clk,
  Rst,
  Load,
  Start,
  LLR,
  Vin0,
  Vin1,
  Vin2,
  Vout0,
  Vout1,
  Vout2,
  BitOut
  );

parameter LLR_WIDTH    = (4)-1;
parameter MSG_WIDTH    = (2)-1;
parameter SIGN_BIT     = LLR_WIDTH;
parameter POS_SAT      = ((1<<(LLR_WIDTH))-1);
parameter NEG_SAT      = -((1<<(LLR_WIDTH)));
parameter ZERO         = 0;

input Clk;                              //Global clock
input Rst;                              //Global reset
input Load;                             //Load LLR data
input Start;                            //Start decoding
input [LLR_WIDTH:0] LLR;                //LLR input
input [MSG_WIDTH:0] Vin0, Vin1, Vin2;   //Ext. message from CN

output [MSG_WIDTH:0] Vout0, Vout1, Vout2;   //Ext. message to CN
output BitOut;                          //Decoded bit out

reg [LLR_WIDTH:0] LLRval;
reg [MSG_WIDTH:0] Vout0, Vout1, Vout2;
reg BitOut;
```

```
wire [LLR_WIDTH:0] AccVal;
wire [MSG_WIDTH:0] Msg0, Msg1, Msg2;

always@(posedge Clk)
begin
  if(Rst)
  begin
    LLRval <= ZERO;
    BitOut <= ZERO;
    Vout0  <= ZERO;
    Vout1  <= ZERO;
    Vout2  <= ZERO;
  end
  else
  begin
    if(Load)
    begin
      LLRval <= LLR;
      Vout0 <= Map4to2(LLR);
      Vout1 <= Map4to2(LLR);
      Vout2 <= Map4to2(LLR);
      BitOut <= ~GetSignBit(LLR);
    end
    else
      if(Start)
      begin
        BitOut <= ~GetSignBit(AccVal);
        Vout0 <= Msg0;
        Vout1 <= Msg1;
        Vout2 <= Msg2;
      end
  end
end

wire [LLR_WIDTH:0] temp;
assign temp = MapAdd3Msg(Vin0, Vin1, Vin2);
assign AccVal = AddSaturate(LLRval, temp);

assign Msg0 = MapAdd2Msg(AccVal, Vin0);
assign Msg1 = MapAdd2Msg(AccVal, Vin1);
assign Msg2 = MapAdd2Msg(AccVal, Vin2);

//Map 4bit message to 2bit
function [MSG_WIDTH:0] Map4to2;
input [LLR_WIDTH:0] Val;
begin
```

```verilog
    case(Val) // Threshold, Tm = 2;
      4'b0000: Map4to2 = 2'b00;
      4'b0001: Map4to2 = 2'b00;
      4'b0010: Map4to2 = 2'b00;
      4'b0011: Map4to2 = 2'b01;
      4'b0100: Map4to2 = 2'b01;
      4'b0101: Map4to2 = 2'b01;
      4'b0110: Map4to2 = 2'b01;
      4'b0111: Map4to2 = 2'b01;
      4'b1000: Map4to2 = 2'b11;
      4'b1001: Map4to2 = 2'b11;
      4'b1010: Map4to2 = 2'b11;
      4'b1011: Map4to2 = 2'b11;
      4'b1100: Map4to2 = 2'b11;
      4'b1101: Map4to2 = 2'b11;
      4'b1110: Map4to2 = 2'b10;
      4'b1111: Map4to2 = 2'b10;
      default: Map4to2 = 2'b00;
    endcase
end
endfunction

function [LLR_WIDTH:0] MapAdd3Msg;
input [MSG_WIDTH:0] Val0, Val1, Val2;
begin
  case({Val0, Val1, Val2})
    6'b000000: MapAdd3Msg = 4'b0011;
    6'b000001: MapAdd3Msg = 4'b0101;
    6'b000010: MapAdd3Msg = 4'b0001;
    6'b000011: MapAdd3Msg = 4'b1111;
    6'b000100: MapAdd3Msg = 4'b0101;
    6'b000101: MapAdd3Msg = 4'b0111;
    6'b000110: MapAdd3Msg = 4'b0011;
    6'b000111: MapAdd3Msg = 4'b0001;
    6'b001000: MapAdd3Msg = 4'b0001;
    6'b001001: MapAdd3Msg = 4'b0011;
    6'b001010: MapAdd3Msg = 4'b1111;
    6'b001011: MapAdd3Msg = 4'b1101;
    6'b001100: MapAdd3Msg = 4'b1111;
    6'b001101: MapAdd3Msg = 4'b0001;
    6'b001110: MapAdd3Msg = 4'b1101;
    6'b001111: MapAdd3Msg = 4'b1011;
    6'b010000: MapAdd3Msg = 4'b0101;
    6'b010001: MapAdd3Msg = 4'b0111;
```

```
6'b010010: MapAdd3Msg = 4'b0011;
6'b010011: MapAdd3Msg = 4'b0001;
6'b010100: MapAdd3Msg = 4'b0111;
6'b010101: MapAdd3Msg = 4'b0111;
6'b010110: MapAdd3Msg = 4'b0101;
6'b010111: MapAdd3Msg = 4'b0011;
6'b011000: MapAdd3Msg = 4'b0011;
6'b011001: MapAdd3Msg = 4'b0101;
6'b011010: MapAdd3Msg = 4'b0001;
6'b011011: MapAdd3Msg = 4'b1111;
6'b011100: MapAdd3Msg = 4'b0001;
6'b011101: MapAdd3Msg = 4'b0011;
6'b011110: MapAdd3Msg = 4'b1111;
6'b011111: MapAdd3Msg = 4'b1101;
6'b100000: MapAdd3Msg = 4'b0001;
6'b100001: MapAdd3Msg = 4'b0011;
6'b100010: MapAdd3Msg = 4'b1111;
6'b100011: MapAdd3Msg = 4'b1101;
6'b100100: MapAdd3Msg = 4'b0011;
6'b100101: MapAdd3Msg = 4'b0101;
6'b100110: MapAdd3Msg = 4'b0001;
6'b100111: MapAdd3Msg = 4'b1111;
6'b101000: MapAdd3Msg = 4'b1111;
6'b101001: MapAdd3Msg = 4'b0001;
6'b101010: MapAdd3Msg = 4'b1101;
6'b101011: MapAdd3Msg = 4'b1011;
6'b101100: MapAdd3Msg = 4'b1101;
6'b101101: MapAdd3Msg = 4'b1111;
6'b101110: MapAdd3Msg = 4'b1011;
6'b101111: MapAdd3Msg = 4'b1001;
6'b110000: MapAdd3Msg = 4'b1111;
6'b110001: MapAdd3Msg = 4'b0001;
6'b110010: MapAdd3Msg = 4'b1101;
6'b110011: MapAdd3Msg = 4'b1011;
6'b110100: MapAdd3Msg = 4'b0001;
6'b110101: MapAdd3Msg = 4'b0011;
6'b110110: MapAdd3Msg = 4'b1111;
6'b110111: MapAdd3Msg = 4'b1101;
6'b111000: MapAdd3Msg = 4'b1101;
6'b111001: MapAdd3Msg = 4'b1111;
6'b111010: MapAdd3Msg = 4'b1011;
6'b111011: MapAdd3Msg = 4'b1001;
6'b111100: MapAdd3Msg = 4'b1011;
6'b111101: MapAdd3Msg = 4'b1101;
```

```
    6'b111110: MapAdd3Msg = 4'b1001;
    6'b111111: MapAdd3Msg = 4'b1000;
    default: MapAdd3Msg = 4'b0000;
  endcase
end
endfunction

//Map and Add 2 messages
function [MSG_WIDTH:0] MapAdd2Msg;
input [LLR_WIDTH:0] Acc;
input [MSG_WIDTH:0] Val;
begin
  case({Acc,Val})
    6'b100000: MapAdd2Msg = 2'b11;
    6'b100001: MapAdd2Msg = 2'b11;
    6'b100010: MapAdd2Msg = 2'b11;
    6'b100011: MapAdd2Msg = 2'b11;
    6'b100100: MapAdd2Msg = 2'b11;
    6'b100101: MapAdd2Msg = 2'b11;
    6'b100110: MapAdd2Msg = 2'b11;
    6'b100111: MapAdd2Msg = 2'b11;
    6'b101000: MapAdd2Msg = 2'b11;
    6'b101001: MapAdd2Msg = 2'b11;
    6'b101010: MapAdd2Msg = 2'b11;
    6'b101011: MapAdd2Msg = 2'b11;
    6'b101100: MapAdd2Msg = 2'b11;
    6'b101101: MapAdd2Msg = 2'b11;
    6'b101110: MapAdd2Msg = 2'b11;
    6'b101111: MapAdd2Msg = 2'b10;
    6'b110000: MapAdd2Msg = 2'b11;
    6'b110001: MapAdd2Msg = 2'b11;
    6'b110010: MapAdd2Msg = 2'b11;
    6'b110011: MapAdd2Msg = 2'b10;
    6'b110100: MapAdd2Msg = 2'b11;
    6'b110101: MapAdd2Msg = 2'b11;
    6'b110110: MapAdd2Msg = 2'b10;
    6'b110111: MapAdd2Msg = 2'b00;
    6'b111000: MapAdd2Msg = 2'b11;
    6'b111001: MapAdd2Msg = 2'b11;
    6'b111010: MapAdd2Msg = 2'b10;
    6'b111011: MapAdd2Msg = 2'b00;
    6'b111100: MapAdd2Msg = 2'b10;
    6'b111101: MapAdd2Msg = 2'b11;
    6'b111110: MapAdd2Msg = 2'b00;
```

```verilog
      6'b111111: MapAdd2Msg = 2'b00;
      6'b000000: MapAdd2Msg = 2'b10;
      6'b000001: MapAdd2Msg = 2'b11;
      6'b000010: MapAdd2Msg = 2'b00;
      6'b000011: MapAdd2Msg = 2'b01;
      6'b000100: MapAdd2Msg = 2'b00;
      6'b000101: MapAdd2Msg = 2'b10;
      6'b000110: MapAdd2Msg = 2'b00;
      6'b000111: MapAdd2Msg = 2'b01;
      6'b001000: MapAdd2Msg = 2'b00;
      6'b001001: MapAdd2Msg = 2'b10;
      6'b001010: MapAdd2Msg = 2'b01;
      6'b001011: MapAdd2Msg = 2'b01;
      6'b001100: MapAdd2Msg = 2'b00;
      6'b001101: MapAdd2Msg = 2'b00;
      6'b001110: MapAdd2Msg = 2'b01;
      6'b001111: MapAdd2Msg = 2'b01;
      6'b010000: MapAdd2Msg = 2'b01;
      6'b010001: MapAdd2Msg = 2'b00;
      6'b010010: MapAdd2Msg = 2'b01;
      6'b010011: MapAdd2Msg = 2'b01;
      6'b010100: MapAdd2Msg = 2'b01;
      6'b010101: MapAdd2Msg = 2'b00;
      6'b010110: MapAdd2Msg = 2'b01;
      6'b010111: MapAdd2Msg = 2'b01;
      6'b011000: MapAdd2Msg = 2'b01;
      6'b011001: MapAdd2Msg = 2'b01;
      6'b011010: MapAdd2Msg = 2'b01;
      6'b011011: MapAdd2Msg = 2'b01;
      6'b011100: MapAdd2Msg = 2'b01;
      6'b011101: MapAdd2Msg = 2'b01;
      6'b011110: MapAdd2Msg = 2'b01;
      6'b011111: MapAdd2Msg = 2'b01;
      default: MapAdd2Msg = 2'b00;
    endcase
  end
endfunction

//Add 4bit values with saturation condition
function [LLR_WIDTH:0] AddSaturate;
input [LLR_WIDTH:0] LLRmsg;
input [LLR_WIDTH:0] MAPmsg;
reg [LLR_WIDTH:0] AccSum;
reg Carry;
reg OverFlowNeg, OverFlowPos;
```

```verilog
begin
  {Carry, AccSum} = LLRmsg + MAPmsg;
  OverFlowNeg = LLRmsg[SIGN_BIT] & MAPmsg[SIGN_BIT];
  OverFlowPos = LLRmsg[SIGN_BIT] | MAPmsg[SIGN_BIT];
  AddSaturate = (Carry & ~AccSum[SIGN_BIT] & OverFlowNeg)? NEG_SAT :
                ((~Carry & AccSum[SIGN_BIT] & ~OverFlowPos)? POS_SAT :
                AccSum);
end
endfunction

function GetSignBit;
input [LLR_WIDTH:0] DataIn;
begin
  casex(DataIn)
    4'b1xxx: GetSignBit = 1;
    4'b0xxx: GetSignBit = 0;
    default: GetSignBit = 0;
  endcase
end
endfunction

endmodule

//////////////////////////////////////////////////////////////
// Module Name     :    Cnode_MMS2
// Designer        :    Vikram A. Chandrasetty
// Email           :    vikramac@ieee.org
// Program         :    Verilog HDL
//
// Description     :
// Verilog HDL module for LUT based Check node (MMS2 algorithm)
//
// Copyright (C) 2017, www.vikramac.com
//////////////////////////////////////////////////////////////

`timescale 1ns / 1ps
module Cnode_MMS2(
  In0,
  In1,
  In2,
  In3,
  In4,
  In5,
  Out0,
  Out1,
  Out2,
  Out3,
```

```verilog
    Out4,
    Out5,
    ParityOut
    );

input  [1:0]In0;   //Ext. message from VN [0]
input  [1:0]In1;   //Ext. message from VN [1]
input  [1:0]In2;   //Ext. message from VN [2]
input  [1:0]In3;   //Ext. message from VN [3]
input  [1:0]In4;   //Ext. message from VN [4]
input  [1:0]In5;   //Ext. message from VN [5]

output [1:0]Out0;   //Ext. message to VN [0]
output [1:0]Out1;   //Ext. message to VN [1]
output [1:0]Out2;   //Ext. message to VN [2]
output [1:0]Out3;   //Ext. message to VN [3]
output [1:0]Out4;   //Ext. message to VN [4]
output [1:0]Out5;   //Ext. message to VN [5]
output ParityOut;   //Parity check output

//Parity Checks
assign ParityOut = In0[1] ^ In1[1] ^ In2[1] ^ In3[1] ^ In4[1] ^ In5[1];
assign Out0[1] = ParityOut ^ In0[1];
assign Out1[1] = ParityOut ^ In1[1];
assign Out2[1] = ParityOut ^ In2[1];
assign Out3[1] = ParityOut ^ In3[1];
assign Out4[1] = ParityOut ^ In4[1];
assign Out5[1] = ParityOut ^ In5[1];

//Input Minimums
assign Out0[0] = In1[0] & In2[0] & In3[0] & In4[0] & In5[0];
assign Out1[0] = In0[0] & In2[0] & In3[0] & In4[0] & In5[0];
assign Out2[0] = In0[0] & In1[0] & In3[0] & In4[0] & In5[0];
assign Out3[0] = In0[0] & In1[0] & In2[0] & In4[0] & In5[0];
assign Out4[0] = In0[0] & In1[0] & In2[0] & In3[0] & In5[0];
assign Out5[0] = In0[0] & In1[0] & In2[0] & In3[0] & In4[0];
endmodule

////////////////////////////////////////////////////////////////
// Module Name    :  RS232_TranRec
// Designer       :  Vikram A. Chandrasetty
// Email          :  vikramac@ieee.org
// Program        :  Verilog HDL
//
// Description    :  Verilog HDL module for RS232 transceiver.
//
// Copyright (C) 2017, www.vikramac.com
////////////////////////////////////////////////////////////////
```

```verilog
`timescale 1ns / 1ps
module RS232_TranRec(
   Clock,
   Reset,
   RxIn,
   RxData,
   RxDone,
   TxOut,
   TxData,
   TxDone,
   TxReady
   );

parameter RS232_DATA_WIDTH    = (8)-1;
parameter CLOCK_FREQ          = 100000000; //100MHz
parameter BAUD_RATE           = 115200;

input Clock;
input Reset;

//Receiver ports
input RxIn;
output RxDone;
output [RS232_DATA_WIDTH:0] RxData;

//Transmitter ports
input TxReady;
input [RS232_DATA_WIDTH:0] TxData;
output TxOut;
output TxDone;

//RS232 Receiver
RS232_Receiver #(CLOCK_FREQ, BAUD_RATE) I_RS232_Rx(
   .clock(Clock),
   .reset(Reset),
   .rx_receiver(RxIn),
   .rx_dataout_ready(RxDone),
   .rx_dataout(RxData)
);

//RS232 Trasmitter
RS232_Transmitter #(CLOCK_FREQ, BAUD_RATE) I_RS232_Tx(
   .clock(Clock),
   .reset(Reset),
   .tx_datain_ready(TxReady),
   .tx_datain(TxData),
   .tx_transmitter(TxOut),
   .tx_transmitter_ack(TxDone)
);
endmodule
```

Sample Verilog HDL codes for implementation of partially-parallel LDPC decoder architecture

```
///////////////////////////////////////////////////////////////
// Module Name   :  DUT_LDPC_PPA
// Designer      :  Vikram A. Chandrasetty
// Email         :  vikramac@ieee.org
// Program       :  Verilog HDL
//
// Description   :
// Verilog HDL top module for LDPC partially-parallel Decoder.
// 1/2 rate (3,6) 2304-bit LDPC code for LTE 4G/5G applications.
// Parallel node factor (P) is 16.  RS232 transceiver is
// integrated in the design for serial communcation interface
// with the decoder.
//
// Copyright (C) 2017, www.vikramac.com
///////////////////////////////////////////////////////////////

`timescale 1ns / 1ps
module DUT_LDPC_PPA(
    CLOCK,
    RESET,
    RS232_DATA_IN,
    RS232_DATA_OUT,
    LDPC_DEC_STATUS,
    LDPC_SOFT_STATUS
  );

//RS232 Interface Parameters
parameter CLOCK_FREQ        = 100000000;   //100MHz global clock
parameter BAUD_RATE         = 115200;      //Baud rate of RS232
                                           transceiver
parameter RS232_DATA_WIDTH  = 8;           //RS232 input and output
                                           data width
```

```verilog
//LDPC Decoder Parameters
parameter LDPC_CODE_LENGTH      = 2304;//Codelength of LDPC decoder
parameter LDPC_PARALLEL_NODES   = 16; //Num. of parallel nodes- check/
variable
parameter LLR_PRECISION         = 4;  //LLR quantization
parameter EXT_MSG_WIDTH         = 2;  //Ext. message: 2-bit for MMS
                                  algorithm
parameter VNODE_DEGREE          = 3;  //Degree of variable node
parameter CNODE_DEGREE          = 6;  //Degree of check node

//IMP: Donot edit these parameters
parameter DATA_FRAME_LEN        = LDPC_PARALLEL_NODES;
parameter BASE_CODE_LEN         = LDPC_CODE_LENGTH;
parameter BASE_EXP_FACTOR       = 1;
parameter RAND_MAT_SIZE         = LDPC_CODE_LENGTH/(CNODE_DEGREE*
                                  LDPC_PARALLEL_NODES);
parameter DEC_ITER_WIDTH        = 4;
parameter BASE_POS_WIDTH        = 3; //Based on NODE_DEGREE (MAX)
parameter RAND_POS_WIDTH        = 6; //Based on BASE_EXP_FACTOR
parameter CYCL_POS_WIDTH        = 5; //Based on RAND_MAT_SIZE
parameter DATA_FRAME_WIDTH      = 5; //Based on DATA_FRAME_LEN
parameter LLR_DATA_WIDTH        = LLR_PRECISION*DATA_FRAME_LEN;

//Ports
input CLOCK;                // Global clock to the design
input RESET;               // Global reset to the design
input RS232_DATA_IN;       // Serial data input to the design
output RS232_DATA_OUT;     // Serial data output from the design
output LDPC_DEC_STATUS;    // Indicates status of the decoder
output LDPC_SOFT_STATUS;   // Indicates status of serial data loading

wire wLoad;
wire wStart;
wire [LLR_PRECISION-1:0]wLLRin;
wire [DEC_ITER_WIDTH-1:0]wMaxIter;
wire [DEC_ITER_WIDTH-1:0]wIterCnt;
wire [DATA_FRAME_LEN-1:0]wDecFrame;
wire [RS232_DATA_WIDTH-1:0]wRxData;
wire [RS232_DATA_WIDTH-1:0]wTxDecData;
wire wRxDataReady;
wire wTxDataReady;
wire wTxDataAck;
wire wDecReady;
wire wDecStatus;
wire wDecReset;
wire wSoftReset;
wire wSoftStatus;
```

```verilog
// Unit for processing serial data to feed into
// the LDPC decoder in appropriate format
INPUT_PROC_UNIT
    #(LLR_PRECISION, RS232_DATA_WIDTH)
    u_input_proc(
    .Clk(CLOCK),
    .Rst(RESET),
    .SerialDataReady(wRxDataReady),
    .SerialDataIn(wRxData),
    .SoftReset(wSoftReset),
    .Load(wLoad),
    .Start(wStart),
    .LLR(wLLRin),
    .MaxIter(wMaxIter),
    .SoftStatus(wSoftStatus)
  );

assign LDPC_SOFT_STATUS = wSoftStatus;

// Parially-parallel LDPC decoder
LDPC_DECODER
  #(DEC_ITER_WIDTH,  LLR_PRECISION, EXT_MSG_WIDTH, VNODE_DEGREE,
  CNODE_DEGREE,  DATA_FRAME_LEN, BASE_CODE_LEN, BASE_EXP_FACTOR,
  RAND_MAT_SIZE,  BASE_POS_WIDTH,  RAND_POS_WIDTH,  CYCL_POS_WIDTH,
  DATA_FRAME_WIDTH)
  u_ldpc_dec(
      .Clock(CLOCK),
      .Reset(wDecReset),
      .Load(wLoad),
      .Start(wStart),
      .MaxIter(wMaxIter),
      .LLRin(wLLRin),
      .IterCnt(wIterCnt),
      .DecFrame(wDecFrame),
      .DecReady(wDecReady)
  );

assign wDecReset = RESET | wSoftReset;

// Unit for processing parallel decoded data from
// LDPC decoder to serial output
OUTPUT_PROC_UNIT
  #(BASE_CODE_LEN, DATA_FRAME_LEN, DEC_ITER_WIDTH, RS232_DATA_WIDTH)
  u_output_proc(
      .Clk(CLOCK),
      .Rst(wDecReset),
```

```
            .DecReady(wDecReady),
            .IterCnt(wIterCnt),
            .DecFrame(wDecFrame),
            .TxDataAck(wTxDataAck),
            .TxDataReady(wTxDataReady),
            .TxDecData(wTxDecData),
            .DecStatus(wDecStatus)
        );

// RS232 serial data transmitter/receiver
RS232_TranRec
    #(RS232_DATA_WIDTH, CLOCK_FREQ, BAUD_RATE)
    u_rs232_t_r (
            .Clock(CLOCK),
            .Reset(RESET),
            .RxIn(RS232_DATA_IN),
            .RxDone(wRxDataReady),
            .RxData(wRxData),
            .TxData(wTxDecData),
            .TxReady(wTxDataReady),
            .TxOut(RS232_DATA_OUT),
            .TxDone(wTxDataAck)
        );

assign LDPC_DEC_STATUS = wDecStatus;

endmodule

/////////////////////////////////////////////////////////////
// Module Name    :  LDPC_DECODER
// Designer       :  Vikram A. Chandrasetty
// Email          :  vikramac@ieee.org
// Program        :  Verilog HDL
//
// Description    :
// Verilog HDL module for partially-parallel LDPC decoder.
//
// Copyright (C) 2017, www.vikramac.com
/////////////////////////////////////////////////////////////

`timescale 1ns / 1ps
module LDPC_DECODER(
        Clock,
        Reset,
        Load,
        Start,
```

```
      MaxIter,
      LLRin,
      IterCnt,
      DecFrame,
      DecReady
    );

parameter DEC_ITER_WIDTH     = 4;
parameter LLR_PRECISION      = 4;
parameter EXT_MSG_WIDTH      = 2;
parameter VNODE_DEGREE       = 3;
parameter CNODE_DEGREE       = 6;
parameter DATA_FRAME_LEN     = 16;
parameter BASE_CODE_LEN      = 2304;
parameter BASE_EXP_FACTOR    = 1;
parameter RAND_MAT_SIZE      = 24;

parameter BASE_POS_WIDTH     = 3; //Based on NODE_DEGREE (MAX)
parameter RAND_POS_WIDTH     = 6; //Based on BASE_EXP_FACTOR
parameter CYCL_POS_WIDTH     = 5; //Based on RAND_MAT_SIZE
parameter DATA_FRAME_WIDTH   = 5; //Based on DATA_FRAME_LEN
parameter LLR_DATA_WIDTH     = LLR_PRECISION*DATA_FRAME_LEN;

//Ports
input Clock;                               //Global clock
input Reset;                               //Global reset
input Load;                                //Load LLR data
input Start;                               //Start decoding
input [DEC_ITER_WIDTH-1:0]MaxIter;         //Set Max iterations
input [LLR_PRECISION-1:0]LLRin;            //LLR data input
output [DEC_ITER_WIDTH-1:0]IterCnt;        //Early exit iter count
output [DATA_FRAME_LEN-1:0]DecFrame;       //Decoded output
output DecReady;                           //Decoder ready status

//Wires
wire wLoadLLR;
wire wVnpInit;
wire wVnpStart;
wire wVnpUpdt;
wire wCnpStart;
wire wCnpUpdt;
wire wParityCheck;
wire [LLR_DATA_WIDTH-1:0]wLLRframe;
wire [DATA_FRAME_LEN-1:0]wDecData;
```

```verilog
// FSM to control loading and starting of LDPC decoding
FSM_CTRL_UNIT
  #(DEC_ITER_WIDTH,   LLR_PRECISION,   EXT_MSG_WIDTH,   VNODE_DEGREE,
  CNODE_DEGREE,   DATA_FRAME_LEN,   BASE_CODE_LEN,   BASE_EXP_FACTOR,
  RAND_MAT_SIZE,   BASE_POS_WIDTH,   RAND_POS_WIDTH,   CYCL_POS_WIDTH,
  DATA_FRAME_WIDTH)
    DecCntrl(
        .Clk(Clock),
        .Rst(Reset),
        .Load(Load),
        .Start(Start),
        .LLRin(LLRin),
        .MaxIter(MaxIter),
        .ParityCheckIn(wParityCheck),
        .DecDataIn(wDecData),
        .LlrLoad(wLoadLLR),
        .VnpInit(wVnpInit),
        .VnpStart(wVnpStart),
        .VnpMsgUpdt(wVnpUpdt),
        .CnpStart(wCnpStart),
        .CnpMsgUpdt(wCnpUpdt),
        .LLRFrameOut(wLLRframe),
        .IterCntOut(IterCnt),
        .DecFrameOut(DecFrame),
        .DecReady(DecReady)
        );

// Unit for loading LLR, decoding and sending the decoded output
PROC_MEM_UNIT
  #(LLR_PRECISION, EXT_MSG_WIDTH, VNODE_DEGREE, CNODE_DEGREE,
  DATA_FRAME_LEN, BASE_CODE_LEN, BASE_EXP_FACTOR, RAND_MAT_SIZE,
  BASE_POS_WIDTH, RAND_POS_WIDTH, CYCL_POS_WIDTH)
    DecProc(
        .Clk(Clock),
        .Rst(Reset),
        .LoadLLR(wLoadLLR),
        .FetchLLR(wVnpUpdt),
        .LLRin(wLLRframe),
        .VnpInit(wVnpInit),
        .VnpStart(wVnpStart),
        .VnpMsgUpdt(wVnpUpdt),
        .CnpStart(wCnpStart),
        .CnpMsgUpdt(wCnpUpdt),
        .DecodedBits(wDecData),
        .ParityCheck(wParityCheck)
        );
endmodule
```

```
////////////////////////////////////////////////////////////////
// ************* Autogenerated File - DO NOT MODIFY *************
// Module Name: LDPC_Net_SP_VNP16_MMS2
//
// Description: Variable Node processing units
//
////////////////////////////////////////////////////////////////
`timescale 1ns / 1ps
module LDPC_Net_SP_VNP16_MMS2 (
    InitState,
    LLRin,
    FromVnpMem,
    FromVnpNode,
    DecodedOut
    );

parameter PARALLEL_FACTOR = (16)-1;
parameter LLR_INPUT_WIDTH = (64)-1;
parameter EXT_MSG_WIDTH = (96)-1;

input InitState;
input [LLR_INPUT_WIDTH:0] LLRin;
input [EXT_MSG_WIDTH:0] FromVnpMem;

output [EXT_MSG_WIDTH:0] FromVnpNode;
output [PARALLEL_FACTOR:0] DecodedOut;

wire [EXT_MSG_WIDTH:0] ToVNP;
wire [EXT_MSG_WIDTH:0] ToMem;

assign ToVNP = {
    FromVnpMem[ 95: 94], FromVnpMem[ 63: 62], FromVnpMem[ 31: 30],
    FromVnpMem[ 93: 92], FromVnpMem[ 61: 60], FromVnpMem[ 29: 28],
    FromVnpMem[ 91: 90], FromVnpMem[ 59: 58], FromVnpMem[ 27: 26],
    FromVnpMem[ 89: 88], FromVnpMem[ 57: 56], FromVnpMem[ 25: 24],
    FromVnpMem[ 87: 86], FromVnpMem[ 55: 54], FromVnpMem[ 23: 22],
    FromVnpMem[ 85: 84], FromVnpMem[ 53: 52], FromVnpMem[ 21: 20],
    FromVnpMem[ 83: 82], FromVnpMem[ 51: 50], FromVnpMem[ 19: 18],
    FromVnpMem[ 81: 80], FromVnpMem[ 49: 48], FromVnpMem[ 17: 16],
    FromVnpMem[ 79: 78], FromVnpMem[ 47: 46], FromVnpMem[ 15: 14],
    FromVnpMem[ 77: 76], FromVnpMem[ 45: 44], FromVnpMem[ 13: 12],
    FromVnpMem[ 75: 74], FromVnpMem[ 43: 42], FromVnpMem[ 11: 10],
    FromVnpMem[ 73: 72], FromVnpMem[ 41: 40], FromVnpMem[ 9: 8],
    FromVnpMem[ 71: 70], FromVnpMem[ 39: 38], FromVnpMem[ 7: 6],
    FromVnpMem[ 69: 68], FromVnpMem[ 37: 36], FromVnpMem[ 5: 4],
    FromVnpMem[ 67: 66], FromVnpMem[ 35: 34], FromVnpMem[ 3: 2],
    FromVnpMem[ 65: 64], FromVnpMem[ 33: 32], FromVnpMem[ 1: 0]};
```

```
Vnode_MMS2 VN0  (.Init(InitState), .LLR(LLRin[ 3: 0]), .MSGin(ToVNP
[ 5: 0]), .MSGout(ToMem[ 5: 0]), .BitOut(DecodedOut[ 0]) );
Vnode_MMS2 VN1  (.Init(InitState), .LLR(LLRin[ 7: 4]), .MSGin(ToVNP
[ 11: 6]), .MSGout(ToMem[ 11: 6]), .BitOut(DecodedOut[ 1]) );
Vnode_MMS2 VN2  (.Init(InitState), .LLR(LLRin[ 11: 8]), .MSGin(ToVNP
[ 17: 12]), .MSGout(ToMem[ 17: 12]), .BitOut(DecodedOut[ 2]) );
Vnode_MMS2 VN3  (.Init(InitState), .LLR(LLRin[ 15: 12]), .MSGin(ToVNP
[ 23: 18]), .MSGout(ToMem[ 23: 18]), .BitOut(DecodedOut[ 3]) );
Vnode_MMS2 VN4  (.Init(InitState), .LLR(LLRin[ 19: 16]), .MSGin(ToVNP
[ 29: 24]), .MSGout(ToMem[ 29: 24]), .BitOut(DecodedOut[ 4]) );
Vnode_MMS2 VN5  (.Init(InitState), .LLR(LLRin[ 23: 20]), .MSGin(ToVNP
[ 35: 30]), .MSGout(ToMem[ 35: 30]), .BitOut(DecodedOut[ 5]) );
Vnode_MMS2 VN6  (.Init(InitState), .LLR(LLRin[ 27: 24]), .MSGin(ToVNP
[ 41: 36]), .MSGout(ToMem[ 41: 36]), .BitOut(DecodedOut[ 6]) );
Vnode_MMS2 VN7  (.Init(InitState), .LLR(LLRin[ 31: 28]), .MSGin(ToVNP
[ 47: 42]), .MSGout(ToMem[ 47: 42]), .BitOut(DecodedOut[ 7]) );
Vnode_MMS2 VN8  (.Init(InitState), .LLR(LLRin[ 35: 32]), .MSGin(ToVNP
[ 53: 48]), .MSGout(ToMem[ 53: 48]), .BitOut(DecodedOut[ 8]) );
Vnode_MMS2 VN9  (.Init(InitState), .LLR(LLRin[ 39: 36]), .MSGin(ToVNP
[ 59: 54]), .MSGout(ToMem[ 59: 54]), .BitOut(DecodedOut[ 9]) );
Vnode_MMS2 VN10  (.Init(InitState), .LLR(LLRin[ 43: 40]), .MSGin(ToVNP
[ 65: 60]), .MSGout(ToMem[ 65: 60]), .BitOut(DecodedOut[ 10]) );
Vnode_MMS2 VN11  (.Init(InitState), .LLR(LLRin[ 47: 44]), .MSGin(ToVNP
[ 71: 66]), .MSGout(ToMem[ 71: 66]), .BitOut(DecodedOut[ 11]) );
Vnode_MMS2 VN12  (.Init(InitState), .LLR(LLRin[ 51: 48]), .MSGin(ToVNP
[ 77: 72]), .MSGout(ToMem[ 77: 72]), .BitOut(DecodedOut[ 12]) );
Vnode_MMS2 VN13  (.Init(InitState), .LLR(LLRin[ 55: 52]), .MSGin(ToVNP
[ 83: 78]), .MSGout(ToMem[ 83: 78]), .BitOut(DecodedOut[ 13]) );
Vnode_MMS2 VN14  (.Init(InitState), .LLR(LLRin[ 59: 56]), .MSGin(ToVNP
[ 89: 84]), .MSGout(ToMem[ 89: 84]), .BitOut(DecodedOut[ 14]) );
Vnode_MMS2 VN15  (.Init(InitState), .LLR(LLRin[ 63: 60]), .MSGin(ToVNP
[ 95: 90]), .MSGout(ToMem[ 95: 90]), .BitOut(DecodedOut[ 15]) );

assign FromVnpNode = {
  ToMem[ 95: 94], ToMem[ 89: 88], ToMem[ 83: 82], ToMem[ 77: 76], ToMem
[ 71: 70], ToMem[ 65: 64],
  ToMem[ 59: 58], ToMem[ 53: 52], ToMem[ 47: 46], ToMem[ 41: 40], ToMem
[ 35: 34], ToMem[ 29: 28],
  ToMem[ 23: 22], ToMem[ 17: 16], ToMem[ 11: 10], ToMem[ 5: 4], ToMem
[ 93: 92], ToMem[ 87: 86],
  ToMem[ 81: 80], ToMem[ 75: 74], ToMem[ 69: 68], ToMem[ 63: 62], ToMem
[ 57: 56], ToMem[ 51: 50],
  ToMem[ 45: 44], ToMem[ 39: 38], ToMem[ 33: 32], ToMem[ 27: 26], ToMem
[ 21: 20], ToMem[ 15: 14],
  ToMem[ 9: 8], ToMem[ 3: 2], ToMem[ 91: 90], ToMem[ 85: 84], ToMem
[ 79: 78], ToMem[ 73: 72],
```

```
   ToMem[ 67: 66], ToMem[ 61: 60], ToMem[ 55: 54], ToMem[ 49: 48], ToMem
[ 43: 42], ToMem[ 37: 36],
   ToMem[ 31: 30], ToMem[ 25: 24], ToMem[ 19: 18], ToMem[ 13: 12], ToMem
[ 7: 6], ToMem[ 1: 0]};
endmodule

//////////////////////////////////////////////////////////////////
// Module Name    : TB_LDPC_PPA
// Designer       : Vikram A. Chandrasetty
// Email          : vikramac@ieee.org
// Program        : Verilog HDL
//
// Description    :
// Testbench for simulating partially-parallel architecture of
// LDPC decoder using RS232 serial interface.
//
// Copyright (C) 2017, www.vikramac.com
//////////////////////////////////////////////////////////////////

`timescale 1ns / 1ps
//Configure
`define TEST_FILE     "TestInput_H2304_MMS.txt"     //LLR input vectors
`define ITER_FILE     "TestIter_H2304_MMS.txt"      //Iteration value
`define VALID_FILE    "TestValid_H2304_MMS.txt"     //Expected values
`define RESULT_FILE   "ResultData_Serial_H2304_
                       MMS.txt"                      //Result dump file

`define DEBUG  0

module TB_LDPC_PPA();

  //Parameters
  parameter CODE_LENGTH          = (2304)-1;
  parameter MAX_ITER_WIDTH       = (4)-1;
  parameter LLR_WIDTH            = (4)-1;
  parameter RS232_DATA_WIDTH     = (8)-1;
  parameter MAX_ITER             = 4'b1010; //10

  //donot alter
  parameter LLR_INPUT_WIDTH = ((LLR_WIDTH + 1)*(CODE_LENGTH + 1))-1;

  //Register bits for RS232
  parameter ITER_HDR = 4'b1100;
  parameter DECO_HDR = 4'b0011;

  //Timing
  parameter CLK   = 10;
  parameter CLK2  = CLK*2;
```

```verilog
parameter CLK4  = CLK*4;
parameter DLY   = CLK/10;
parameter DLY2  = DLY*2;

//States
parameter INITIAL          = 4'b0000;
parameter RESET            = 4'b1000;
parameter CONFIG_ITER      = 4'b0100;
parameter LOAD             = 4'b0010;
parameter START            = 4'b0001;

//Ports
reg tClock;
reg tReset;
reg [RS232_DATA_WIDTH:0] tSerialRegOut;
reg tSerialOutReady;

wire tSerialOutAck;
wire [RS232_DATA_WIDTH:0] tSerialRegIn;
wire tSerialInReady;
wire tDebugReg;

//Intermediate data
wire SerialBitOut;
wire SerialBitIn;
reg      TaskDone;
reg      [LLR_WIDTH:0] LLRin;
reg      [CODE_LENGTH:0] DecodedOut;
reg      [MAX_ITER_WIDTH:0] UsedIter;
reg      [LLR_INPUT_WIDTH:0] llrData;
reg      [CODE_LENGTH:0] validData;
reg      [MAX_ITER_WIDTH:0] validIter;
reg      [15:0]SerialBitCnt;

integer tFile, iFile, vFile, oFile, tData, iData, vData;
integer frameCnt, loop, count, failedCnt, complete,
debug, simStarted;

//Design Under Test
DUT_LDPC_PPA
  dut_ldpc_ppa(
    .CLOCK(tClock),
    .RESET(tReset),
    .RS232_DATA_IN(SerialBitIn),
    .RS232_DATA_OUT(SerialBitOut),
    .LDPC_DEC_STATUS(tDebugReg),
    .LDPC_SOFT_STATUS(),
    );
```

```
//RS232 Transceiver Instance
RS232_TranRec
   I_RS232_TranRec(
        .Clock(tClock),
        .Reset(tReset),
        .RxIn(SerialBitOut),
        .RxData(tSerialRegIn),
        .RxDone(tSerialInReady),
        .TxOut(SerialBitIn),
        .TxData(tSerialRegOut),
        .TxDone(tSerialOutAck),
        .TxReady(tSerialOutReady)
        );

//Generate Clock
initial
begin
   tClock = 1;
   forever #(CLK/2) tClock = !tClock;
end

//Reading Test/Validation data from file
initial
begin: test_parallel

   //Initialization Signals
   simStarted = 0;
   complete = 0;
   frameCnt = 1;

   //Reset the Decoder
   tReset = 1;
   #(CLK) tReset = 0;

   //Configure MaxIterations
   tSerialRegOut = {CONFIG_ITER, MAX_ITER};
   tSerialOutReady = 1;
   #(CLK) tSerialOutReady = 0;

   while(tSerialOutAck)
   begin
      #(CLK) tSerialOutReady = 0;
   end
   $display("Configured Max Iter successfully!!\n");
   if(`DEBUG)
      debug = `DEBUG;
   else
      debug = 1;
```

```verilog
//Open files to read and write data
tFile = $fopen('TEST_FILE, "r");
if (tFile == 0)
  disable test_parallel;

iFile = $fopen('ITER_FILE, "r");
if (iFile == 0)
  disable test_parallel;

vFile = $fopen('VALID_FILE, "r");
if (vFile == 0)
  disable test_parallel;

oFile = $fopen('RESULT_FILE, "w");
if (oFile == 0)
  disable test_parallel;

TaskDone = 1;
//Loop through the file to fetch the data
    while (!$feof(iFile) && !$feof(vFile) && debug)
    begin

    if(TaskDone)
      begin
        TaskDone = 0;
        //Load the decoder with LLR
        tData = $fscanf(tFile, "%x", llrData);
        if(tData != 1)
        begin
          //Close the files
          $fclose(tFile);
          $fclose(iFile);
          $fclose(vFile);

          //Set Simulation complete flag
          complete = 1;
          TaskDone = 1;
        end
        for(loop = 0; loop <= CODE_LENGTH; loop = loop + 1)
        begin
          #(CLK2) LLRin = llrData[LLR_WIDTH:0];
          llrData = llrData >> (LLR_WIDTH + 1);
          tSerialRegOut = {LOAD, LLRin};
          tSerialOutReady = 1;
          #(CLK) tSerialOutReady = 0;
```

```
                    while(tSerialOutAck)
                    begin
                      #(CLK) tSerialOutReady = 0;
                    end
                 end

                 //Start decoding
                 #(CLK4) tSerialRegOut = {START, 4'b0};
                 tSerialOutReady = 1;
                 #(CLK) tSerialOutReady = 0;

                 while(tSerialOutAck)
                 begin
                   #(CLK) tSerialOutReady = 0;
                 end
                 iData = $fscanf(iFile, "%d", validIter);
                 vData = $fscanf(vFile, "%b", validData);

                 if(`DEBUG)
                   debug = debug - 1;
                 frameCnt = frameCnt + 1;
                 simStarted = 1;
              end
           #(CLK2) simStarted = 1;
      end
end

//Receiving decoded data serially
//(Decoded bits and Used iteration count)
always@ (negedge tClock)
begin

   if(tReset | (tSerialRegOut[7:4] == LOAD))
   begin
     DecodedOut = 0;
     UsedIter = 0;
     SerialBitCnt = 0;
   end
   else
   begin
     if(tSerialInReady & !TaskDone)
     begin
       if(SerialBitCnt == CODE_LENGTH + 1) //CodeLength + Iter data
         TaskDone = 1;

       SerialBitCnt = SerialBitCnt +1;
```

```verilog
        case(tSerialRegIn[7:4])
          ITER_HDR:
            begin
              UsedIter = tSerialRegIn[3:0];
            end
          DECO_HDR:
            begin
              DecodedOut = DecodedOut >> 1;
                DecodedOut[CODE_LENGTH] = tSerialRegIn[0];
            end
          default: $display("Something Wrong!!\n");
        endcase
    end
  end
end

//Decoded data validation
always@ (negedge tClock)
begin
  if(tReset)
  begin
    count = 1;
    failedCnt = 0;
  end
  else
  if(TaskDone && simStarted && (count == (frameCnt-1)))
  begin
    if((validData == DecodedOut) && (validIter == UsedIter))
    begin
      $display("Frame: %d => PASSED (Valid: %d Decoded: %d)",
        count, validIter, UsedIter);
      $fwrite(oFile, "Frame: %d (Valid: %d -- Decoded: %d) : PASSED\n",
        count, validIter, UsedIter);
    end
    else
    begin
      failedCnt = failedCnt + 1;
      $display("Frame: %d => FAILED(Valid: %d Decoded: %d)",
        count, validIter, UsedIter);
      $fwrite(oFile, "Frame: %d (Valid: %d -- Decoded: %d) : PASSED\n",
        count, validIter, UsedIter);
    end
    $fflush(oFile);
    count = count + 1;
  end
end
```

```verilog
//Print simulation result
always@ (negedge tClock)
begin
  if(TaskDone && complete)
  begin
    if(failedCnt)
      $display("Simulation FAILED (%d) !!\n", failedCnt);
    else
      $display("Simulation PASSED !!\n");
    //Stop the simulation
    $stop;
    //Close file
    $fclose(oFile);
  end
end

endmodule
```

Index

Printed in the United States
By Bookmasters